Responsonomics

Assessing the Value of IT Incident Management

Rob Schnepp & Ron Vidal

The Amazon Endure typeface was designed by 2K/DENMARK in 2025.
Template id: ST-414D415A-25-A01
Printed in The United States.
ISBN: 979-8-9935654-0-8

DEDICATION

This book is dedicated to all the IT incident responders out there, defending the value of IT systems around the world.

CONTENTS

ACKNOWLEDGMENTS

This textbook introduces Responsonomics as an original framework for applying economic principles to IT incident response. The concepts of financial observability, erosion of customer and investor trust, economic price, labor substitution, and incident cost modeling are developed here as distinct contributions.

Throughout the chapters, references are made to well-known works in reliability engineering, DevOps, and management science (e.g., The DevOps Handbook, Site Reliability Engineering, The Balanced Scorecard). These works are cited to anchor Responsonomics in existing scholarship and to acknowledge the intellectual traditions that inform this field.

The writing in this book is original. Responsonomics builds upon and extends recognized concepts—such as process automation, performance measurement, and decision modeling—by reframing them in economic terms that emphasize cost, opportunity, and organizational value.

The authors wish to thank the experiences and input from thousands of responders trained by Blackrock Partners over the last twelve years. Without them, the concept of Responsonomics would not have been possible.

1 VALUE AND INCIDENT MANAGEMENT

*Only under pressure do systems reveal their true economic value—
or their silent inefficiencies*

Organizations may not fully appreciate their value during periods of stability. When systems behave as expected, performance is easy to mistake for competence, and perhaps a false sense of security sets in simply because of habit. It is only under stress—when systems fail, time compresses, and uncertainty dominates—that value, or the lack thereof, becomes visible, measurable, and consequential.

This book begins from a simple but often overlooked premise: *Incident Management* is not merely a technical discipline—it is an economic one, and each *incident* is an economic event that puts value at risk. Whether the disruption is large or small, the organization immediately begins spending scarce resources when systems are degraded. These resources—such as time and money—are limited in supply and must be allocated strategically to maximize the resolution effort. In Incident Management, scarcity often reveals itself under stress, when multiple priorities compete for the same limited responder attention, system capacity, or leadership focus. Recognizing and managing these constraints is essential to cost-effective, resilient operations. The cost of these constraints accrues regardless of whether they are tracked, reported, or acknowledged. The absence of formal accounting does not mean the absence of economic impact.

Traditional approaches to Incident Management focus heavily on tools, processes, and metrics such as system uptime and *Mean Time to Recover (MTTR)*. While necessary, these measures are incomplete without additional business related context. They describe what happened, but they often fail to identify the costs, or what value was quietly destroyed in the process of recovery. As a result, companies hope for speedy recovery and in some cases declare success even as trust, morale, and long-term resilience erode.

At the center of the Responsonomics framework are three interdependent currencies of time, trust, and cost. These are not abstract concepts; they are assets that depreciate rapidly during incidents and recover slowly afterward. Effective Incident Management does not eliminate the loss or damage to these currencies—no response can—but it determines which losses are minimized, which are accepted, and which are unknowingly compounded.

Importantly, Responsonomics does not argue for the cheapest response, the fastest response, or the easiest response. It argues for the most economical response across the full lifecycle of an incident—from detection through recovery, learning, and prevention. In this view, preparedness is not overhead and process improvement is not optional. Each is an investment decision with measurable economic consequences.

This book is written for executive leaders, all types of responders, engineers, and other decision-makers who operate in IT environments where failure is inevitable, but outcomes are not. It is for those who recognize that incidents are moments of risk—but also moments of clarity. When managed well, they reveal where value truly lies. When managed poorly, they expose the cost of ignoring it. This book is not intended to deep dive into the specifics of incident resolution. Rather is intended to position the acts of trouble shooting and resolution as segments of Responsonomics. For more information and specifics on the roles and responsibilities of incident response, see our other textbook, *Incident Management for Operations* by Rob Schnepp, Ron Vidal, and Chris Hawley. O'Reilly Media, 2017.

Responsonomics provides the language, models, and decision lenses to see Incident Management for what it is: a continuous negotiation between urgency and judgment, speed and stewardship, action and consequence. By understanding the value dynamics at play, organizations can move beyond reactive recovery toward deliberate, economically informed resilience. The chapters that follow are an invitation to stop treating incidents as emergencies—and to start recognizing them as defining moments where value is either protected or permanently lost.

2 INTRODUCTION TO RESPONSONOMICS

Incidents are an existential threat to your business

Today, nearly every type of company depends on IT systems to run its operations and create competitive advantages. Every company that relies on technology for any part of their business is vulnerable to financial loss, reputational risk, erosion of customer and investor confidence, or other adverse outcomes from IT performance degradations, service outages or security incidents. It should be no surprise that customers, vendors, and employees demand always on, highly reliable products and services, nearly anywhere in the world. When IT incidents occur—and they will—organizations must respond quickly and efficiently to resolve the issue, much like a fire department. To that end, efficient incident response to any high severity technology issue, whether it is customer facing or not, underpins the fundamental trust equation between the company and any other organization or individual that relies on it. Incident response then, is solely focused on the act of resolution.

Incident response, however, is not the same thing as Incident Management and it's critical to understand the difference.

Incident Management is a broad, high level, formal discipline that provides the foundation for all incident response related activities. Incident Management is about creating and supporting the underlying processes that allow incident response to move forward with purpose and clarity—before any incident occurs. It is the pre-incident foundation for training the right employees, assembling the right people at the right time to make the right decisions, establishing defined roles and responsibilities, communicating incident updates, and having the right type of executive level support during resolution. Incident Management supports incident response and to a great degree, how efficiently a company resolves is linked to how much time is spent refining the company's Incident Management program.

If Incident Management is viewed holistically, as an integral part of any company culture, it is easy to see the importance of what Sociologist Émile Durkheim termed a system of *organic solidarity*. This occurs when different specialists must cooperate under a set of shared norms, working as a team to reduce the *friction cost* during incidents.

Friction cost is generated by factors such as:

- Poor cross-team collaboration
  (Silos, turf battles, hesitation to escalate, unclear ownership)
- Inefficient communication channels
  (Too many Slack channels, unclear updates, information fragmentation)
- Unclear or shifting roles
  (No clear Incident Commander, SMEs talking over each other, executives swooping in and taking over)
- Too many unnecessary responders
  (Lurkers increase labor and opportunity cost (OC))
- Ineffective tooling or visibility gaps
  (Slow detection, inaccurate alerts, poor dashboards)
- Slow decision-making caused by hierarchy or culture
  (Peacetime rank interfering with wartime urgency)

Each minute of friction increases the economic impact of an incident and ultimately the price the company pays in trust and reputation.

> *Key Thought:* Émile Durkheim (1858–1917) argued that societies depend on shared norms, roles, and cohesion to function effectively. His theories provide a useful lens for understanding the Responsonomics framework in IT incident management.

Friction Kills Operational Efficiency

Friction is the natural enemy of operational maturity—because operationally mature companies can resolve with the right amount of people, at a dramatically lower cost, simply by reducing friction in how teams collaborate. Operational maturity defines how consistently, efficiently, and effectively an organization manages its processes, people, and technology to deliver reliable services, handle risks, and adapt, moving from reactive fixes to proactive, strategic operations, often measured by metrics like incident resolution time and service consistency.

Without solidarity—and unity of purpose— resolution slows, friction increases, and the price tag of incident response swells. Incident Management creates the backdrop and operational rules of engagement so that everyone sees the goals and objectives as one.

Incident response, then, becomes the beneficiary or victim for how effectively Incident Management is accepted by the company. Even a great group of responders will have a tough time being great if they must battle poor company culture, untrusting executive leadership, or any other obstacles that inhibits forward momentum, on top of resolving the actual incident. Incident response also provides feedback into Incident Management after each incident by way of observations, reports, *After Action Reviews (AAR)* and other tools, to refine the process of response and resolution.

> ***Key Thought:*** At the end of the day, how well incident response teams assemble, communicate, document, troubleshoot, resolve, recover, and continuously improve performance has a direct impact on how the incident is viewed and remembered by a company and its stakeholders.

In many cases, elongated resolution times for incidents are a result of many factors that have not been optimized before the incident even starts: ineffective monitoring tools and/or alert thresholds, inefficient processes for assembling the right responders, lack of clarity in the response process, lack of end-to-end responsibility/leadership of the incident, poor cooperation between business units, ineffective communication practices and the inability to learn from past incidents. On top of that, there is a real price to be paid for IT outages and service disruptions—not just in lost revenue dollars, but in things like reputational damage, sales delays in closing future contracts, diversion of employees from duties other than incident response, and missed opportunities to achieve the strategic business goals and objectives of the organization.

Responsonomics offers a different way to look at reducing friction points in Incident Management and incident response within your company. Incident resolution should no longer look like an orphan effort, underfunded, understaffed or under supported by the company. Without getting too esoteric, high functioning and operationally mature response teams behave like a socioeconomic mission critical part of the company, just like sales, product development, or executive leadership.

Responsonomics Overview

Responsonomics provides a different lens to view the process of improving operational maturity by way of finding and quantifying inefficiency and replacing it with predictability. If Incident Management is the foundation of incident response, predictability is the foundation for Incident Management. Responsonomics overlays well established economic concepts onto Incident Management, reframing service outages, performance degradations, and security events not just as technology interruptions—but also as novel microeconomic episodes that impact employees' time (referred to in the rest of the book as *labor*), overall productivity, company reputation, brand value, investor confidence and organizational efficiency. The incidents may be novel, but the company's approach to resolving them can have quite a high level of predictability.

A Responsonomic approach to Incident Management helps target areas of improvement, spotlighting the many direct and indirect social and economic influences on incident resolution. Responsonomics can also assist leaders with making a strong business case for Incident Management improvements using business and economic terms, beyond standard operational metrics. By identifying places in the incident lifecycle (more on this later in the book) where true cost challenges lie, we can understand the actual price to the company of service disruptions and outages, and how specific process improvements benefit the entire company's business. It empowers all involved with incident resolution to be better custodians of that most precious commodity—time— the one resource that is non-renewable for everyone on the planet!

At its core, Responsonomics redefines incident response not simply as a quest for reduction in downtime metrics—it reframes the entire practice as a strategic exercise in resource allocation, business priorities, cost control, and value optimization. Traditional incident response thinking has trained teams to ask, "How fast can we fix technology failures?"—a question rooted in urgency and often driven by an emotional response to the highest severity, highest priority major incidents. While speed remains critically important, Responsonomics introduces a different question:

> "How do we take a value-driven approach to understanding the intricate relationships between all facets of Incident Management that result in efficient incident response?"

When framed this way, speed is not dismissed—but repositioned. It is no longer the primary metric of success. Instead, speed becomes a byproduct of leadership, operational maturity, organizational efficiency, disciplined coordination, and scalable systems. The goal is not just to move faster, but to move smarter—ensuring that every decision made during incident response contributes in some way to long-term resilience, cost-effectiveness, and organizational learning.

> *Key Thought:* Speed does not translate to efficiency, but efficiency will ultimately result in increased speed of resolution.

Those in turn directly impact the bottom line—the price paid for the IT failure. Responsonomics trains responders and leaders to create an operational mindset whereby technical experts become incident responders who have technical talent, not technologists trying to solve an IT issue— a subtle but important difference that helps responders adopt a mindset like this: Fire is not an emergency to the Fire Department—it's what we do!

With Responsonomics, several key outcomes emerge as vital to a mature, high-functioning incident response capability:

- Development of a cost-conscious use of on-call labor (technical subject matter experts (SME) and other resources) by eliminating low-value alerts and repetitive low severity incidents and events.
- Ensuring alignment of incident response doctrine and workflows with *business impact (BI),* ensuring the right incidents are appropriately resourced and resolved with purpose.
- Improvement of internal and external communication protocols to reduce confusion and provide actionable information to key stakeholders away from the primary resolution effort.
- Creation of predictable and repeatable processes for operational feedback, so every incident teaches and improves the system—not just the people.
- Understanding the importance of cost as it relates to the speed of resolution, recognizing that rapid resolution at excessive resource cost is not sustainable, but sometimes necessary.

These strategic shifts are what differentiate Responsonomics from traditional incident response doctrine. It is not just a toolkit for responders and practitioners—it is a philosophical approach implemented to build and understand the business case for intentional response at scale. The remainder of the book will dig into these broad concepts and provide you with a deeper understanding of a Responsonomic mindset, and a road map to improve the operational maturity of your people and processes.

3 MICRECONOMICS APPLIED TO INCIDENT MANAGEMENT

Every incident consumes scarce resources. The question is how wisely you allocate those resources

Responsonomics incorporates four fundamental economic principles into the practice of Incident Management. They are *supply, demand, cost, and price.* These concepts help executive leadership and responders understand the business impacts of incident response, and the real implications of inefficient incident resolution. In short, understanding the basic parameters of Responsonomics will help leaders understand the key influences that impact incident response capabilities.

Each concept is offered to assist in diagnosing, analyzing, and improving how response resources are allocated during incidents. Throughout the remainder of the book, we will demonstrate ways to integrate these concepts into your new perspective on Incident Management and incident response.

Supply: The Pipeline of Incidents

In conventional economics, supply refers to the specific amount of goods and services available to meet demand. There are more specific components to the traditional definition but for our purpose here, it is sufficient to make the point. In the context of Respononomics, supply represents the number and type of incidents and events that enter or have the potential to enter the incident response process.

Although definitions vary by organizations, for the purposes of this textbook an *incident* is an unplanned performance degradation, service outage or security breach that poses an operational, financial, and/or reputational risk to the company. Incidents may or may not be customer impacting, but fundamentally, incidents adversely impact the availability of an IT environment and by extension, the normal operations of the business. Incidents may be of different severity, but each require intervention, specifically matching the right group of trained, domain specific responders to the conditions of the incident, to bring the IT environment back into a normal operational state.

An *event* is anything else that does not pose an acute risk to the IT environment and while it may require immediate attention, it does not pose an operational, financial, and/or reputational risk to the company. Events are non-threatening to internal IT operations, external customer needs, etc. Events don't call for the same sense of urgency as incidents nor is their resolution as time dependent. That is not to say they aren't disruptive in some way, and they must be mitigated, but it's about understanding the impact to the business or its customers.

> *Key Thought:* The supply side of Responsonomics is the very first place to look for overall incident response efficiency and cost reduction. The less supply of incidents, the less demand for incident response, and likely an overall lower total price for responses over time to the company.

Supply refers to the volume, intensity, and timing of incidents that require human or system attention. In a Responsonomics context, supply increases when systems are fragile, deployment velocity is high, changes are poorly controlled or under tested, external influences (traffic spikes, provider outages) introduce complexity, vendor activities create issues, or customer activities impact the company's IT environment. As reported from our global customer base, approximately 70% of incidents are change related or otherwise self-inflicted.

Supply is not static and fluctuates based on many factors that are unique to each type of business. The more incidents a company has, the more demand there is for incident response. Consequently, there is more risk of inefficient resolution simply based on the volume of incidents.

Here, we examine five impactful contributors to supply volume through the lens of Responsonomics. Each of these factors represents a lever of economic control that can either amplify or mitigate supply.

1) System Instability

System instability is a foundational driver of supply. Unstable systems are characterized by frequent alerts, false positives, performance degradations, or full-scale service outages. These disruptions often stem from architectural decisions, defects in code, fragile or non-resilient infrastructure, or insufficient automated testing. Each failure—no matter how minor—places load on the incident response system, diverting the attention of key technical and leadership resources to incidents rather than doing the strategic work of building the business.

From a Responsonomics perspective, system instability behaves like a recurring cost center. Proactive investments in reliability engineering, automated testing, and technical debt reduction are therefore not just quality measures—they're incident response related cost controls.

2) Deployment Velocity

While rapid code deployments are a core principle of IT practices, it also introduces additional vectors for instability. Each deployment—regardless of how small—carries some possibility of failure. IT environments with high-velocity deployment pipelines must balance innovation with resilience. Without adequate mechanisms for monitoring, rollback, and learning, the marginal cost of each deployment can lead to an overall negative return on engineering investment.

In critical infrastructure, such as power systems, pipelines and oil refineries, change is minimized. When a change needs to be implemented, it is meticulously planned and implemented with numerous safeguards. The best way to run critical infrastructure is to get all the settings and configurations just right, then lock the doors and don't let any people near any of the machines. Today's IT environments are the complete opposite!

3) Architectural Complexity

Systems with deeply coupled components, microservices with opaque interactions, and legacy integrations all introduce what is often called *emergent fragility*—failures that cannot be anticipated by examining any single component in isolation.

When novel incidents occur in complex systems, correlating problem to resolvers is more complicated, and establishing a common operational viewpoint via SEV levels is more difficult. Troubleshooting takes longer, and the labor and operational cost per minute of downtime increases sharply. From a Responsonomics standpoint, complexity raises the overall cost of responding to incidents, along with the eventual price tag associated with resolution. Economic mitigation strategies for optimized supply include clear service ownership, dependency mapping, and reducing *Mean Time to Detection (MTTD)* through better observability.

4) External Dependencies

Modern systems rely heavily on third-party services, including cloud infrastructure, network operators, payment processors, customer activities, and external APIs to name a few. These integrations, while powerful, introduce externalized risk: volatility and failure points that originate outside the organization's control. The companion to externalized risk is externalized incident response. In many cases, Internal incident response teams do not own or control the external response—the third-party service provider does. And, depending on how an internal response team is measured, that third-party response can directly influence internal operational metrics.

When an external service fails or violates its *Service Level Agreement (SLA)*, the internal team may have little to no ability to resolve the issue. This creates an asymmetrical cost burden: The company pays the price in downtime, *escalations*, and customer support, while the vendor remains operationally and potentially financially insulated. Responsonomic analysis urges teams to track the cost of dependence volatility and build economic buffers such as contracts that are not just focused on traditional elements (financial terms, SLA, renewal options, etc.), but include more detail on incident response communications, shared resolution responsibility, system redundancy, failover strategies, and vendor competency regarding incident response. At the end of the day, you want all third parties to see incidents the same way you do.

> *Key Thought:* Companies can outsource the operating expense to third parties, but they never outsource the risk

This same concept applies to teams with a decentralized response model. With this approach, smaller product or DevOps teams may have responsibility for initial detection and troubleshooting, potentially escalating to a *Major Incident Management (MIM)* team after some period of time. In some cases, the MIM team is held accountable for the *Mean Time to Resolve/Repair/Restore/Remediate (MTTR)* numbers, even when the decentralized team may have waited too long to escalate or is in some other way inefficient in its own ability to resolve an issue.

5) Alerting Strategy and Burnout

Poorly tuned alerts, low signal-to-noise ratios, or misaligned thresholds generate phantom incidents—alerts that may feel urgent but require no actual intervention. Over time, this creates a false supply of incidents and a false demand for incident response, wasting time and cognitive cycles of the responders and increasing the risk of alert fatigue.

Burnout among responders is not just a human issue—it's an economic liability. As alert volume grows, human attention capacity diminishes and impacts from context switching increases (moving from regular duties to incident response mentality). In Responsonomics, this is modeled as *capacity leakage*—the progressive erosion of usable employee time due to poor signal design. Investing in smarter and up-to-date alerting—via machine learning, *Service Level Objective (SLO)* based thresholds, or responder feedback loops—returns measurable economic value by preserving the time and attention of responders for high-impact incidents.

This is an area where AI Operations can be beneficial in fine tuning the supply side of incident response. While it requires a large amount of upfront effort to set the company up for efficient observability, the investment pays dividends in the long run in terms of addressing burnout and capacity leakage. If incidents can be prevented due to architectural decisions, vendor selection, process improvement, reduction of technical debt, accurate documentation, or other initiatives, the impact on the supply of incidents can be profound.

Even with a strong incident prevention program, incidents will still occur, but reducing the supply should be a top priority. The discussion of supply must also include a nod to the classification of an incident, commonly referred to as *Severity (SEV) levels or Priority (P) levels.* Whatever term you use, it is vitally important to be clear, concise, neither too granular or too broad, and consistent throughout the company.

The concept of *severity levels* in IT incident response has evolved as a crucial mechanism for classifying incidents based on their impact on business operations. This stratification of incidents into severity levels—commonly ranging from SEV-1 (most critical) to SEV-4 (least critical)—is not merely a technical taxonomy, but a reflection of how digital systems intertwine with BI, customer experience, and reputational risk. Despite convergence around a typical four-tier system (SEV-1 to SEV-4), companies often customize definitions based on risk tolerance, customer expectations, and regulatory requirements. In financial services or healthcare, for example, a data integrity issue may be escalated to SEV-1 faster than in a consumer-facing app due to compliance implications.

> *Key Thought:* It is important to acknowledge that SEV levels are merely a naming convention that typically outlines communication protocols, notifications, executive involvement and other particular actions. They should not carry any cultural or internal company political consequence. SEV levels outline the rules of engagement for a specific response. Said differently, there should be no shame or hesitation to declare a major incident, if there is any indication of significant business impact.

Severity levels also contribute to understanding other response oriented metrics and inform post-incident learnings. Referred to by various names such as *Postmortems, Root Cause Analysis (RCA)* or causal factors, they all ask the question, "What are the thing or things that broke in the technology stack?" This question is important to make the technology stack more resilient. Equally important is the AAR, which seeks to identify incident response process improvements in advance of the next incident, in areas such as team assembly, level of preparation and participation of *subject matter experts (SME),* stakeholder communications, troubleshooting and resolution. An AAR asks a different macro level question than a technical postmortem. In the pursuit of improving Incident Management, understanding how efficiently the people responded to the thing or things that broke is equally important as understanding the technology failure.

As such, the classification of an incident is not just a technical decision, it also an operational parameter. The development and implementation of accurate IT severity levels reflect a broader maturity in how organizations deploy incident response resources. From informal rules of thumb to codified frameworks integrated with BI, severity levels are foundational to the rules of engagement for team assembly, troubleshooting, and in many cases, internal and external stakeholder communications. Additionally, as a marker to assist in understanding and refining the supply side of incident response, the establishment, communication, and company-wide adoption of clear and concise SEV levels will help Incident Management leaders better understand the linkage between the supply of incidents and the level of response required to address them. As systems become more distributed and interdependent, the precision and consistency of severity classification will remain critical to effective incident response.

Demand: Matching Time, Tools, and Talent to the Incident

In traditional economics, *demand* refers to the quantity of a good or service that consumers are willing to purchase at a given price. In Respononomics, demand takes on a unique operational meaning: it represents the human and technology resources required to efficiently resolve incidents. To efficiently respond to the supply of incidents, companies must allocate a finite and specific amount of three critical elements: Time, Talent, and Tools!

> *Key Thought:* In the context of Responsonomics, the term demand is intended to represent all types of responders that may arrive to resolve an incident. This could include technical SMEs, senior leadership, support teams, legal representatives, or others.

Any incident, or event for that matter, requires a unique combination of these three elements to mitigate the impact of a technology failure. Regardless of what the impact is, these three influences are key to understanding the true cost of responding to a single incident or a number of incidents over time.

Imbalances between high supply and inadequate demand result in systemic strain: overloaded teams, delayed resolution, increased burnout of individuals, and decreased productivity for non-incident response productivity. Again, if supply is optimized to just those incidents that are truly high severity in nature, the need for time, talent, and tools will be reduced, resulting in an overall price reduction for incident response.

Demand should be measured not only in terms of how many incidents entered the system for resolution, but across multiple dimensions that more accurately describe their relevance to understanding the supply pipeline.

Some examples can be found in Figure.

Dimension	Description
Volume	Number of incidents (Supply) per unit of time (daily/weekly/monthly by shift and/or geographic territory, seasonal trends, etc.)
Severity Mix	Proportion of high severity vs. low severity incidents.
Spikes	Frequency of Supply spikes, often around deployments, software upgrades or other scheduled business events (such as Cyber Monday for ecommerce companies or worldwide user conferences for platform companies)
Labor Demand	Average numbers of personnel required per incident + time committed to the incident.
Predictability	Ability to forecast demand based on historical patterns

Figure 3.1

Basic Demand Modeling

Just as retail businesses forecast customer demand to optimize inventory levels and avoid understock, backorders, or overproduction, incident response teams must develop a disciplined understanding of incident supply—the volume, frequency, and severity of incidents that require human intervention. Without this forecasting capability, organizations are reacting—not responding with purpose and a level of planning—unable to routinely align the right resources with the right problems at the right time.

> *Key Thought:* Many teams rely on chronically outdated, manually curated, or ad-hoc on-call schedules—systems that often fail under the pressure of high-severity incidents. When responders are unavailable, delayed, or improperly assigned, every downstream activity—triage, mitigation, communication, and resolution—is compromised. On-call rotations should be built with a plan to keep them current.

Response demand forecasting is foundational to optimized resource allocation. It informs not only staffing models, but also the design and deployment of automation systems, AI-driven alerting, and on-call rotations for SMEs. In Responsonomics, labor availability is a constraint, and constraints must be managed with the same rigor as any other scarce resource. While it is true that incidents are unplanned by nature, operationally mature Incident Management programs continuously collect and analyze incident data to spot trends that inform supply. Without a clear understanding of how often, how long, and how intensely your team is needed, you cannot staff intelligently, manage costs, automate effectively, or justify strategic investments.

Although an imperfect science, modeling incident patterns—by type, time of day, BI, affected technology elements, incident duration etc. will reveal valuable insights—allowing teams to assemble faster with a better match between responders and the type of problem; a foundational aspect of managing the cost of human resources.

Some of the key strategic uses of demand modeling in Respononomics include:

- Aligning on-call coverage with peak supply periods.
- Reducing unnecessary delays and resource burnout due to inefficient call out procedures for responders.
- Adopting a company-wide philosophy that incident response is not a spectator sport. Dispatched resources are those who have a specific set of skills required to move a novel incident forward toward resolution. Everyone that arrives at the incident should be a key contributor, intentionally dispatched by functional role based on the needs of the incident, not curiously or silently lurking in the shadows. Anyone other than a dispatched resource resides in the *notification* pool—no call to action, just being kept in the loop. Companies can notify the world of an incident, but their incident responders should be intentionally

dispatched and expected to respond urgently to an incident channel.

- Learning from incident supply to optimize demand.
- Estimating total labor cost across demand patterns.

These are all strong elements of a business case that can be used to defend your Incident Management budget or make plans to expand the incident response team. By putting good metrics and modelling behind these elements, leaders can approach change intelligently and with facts.

Cost: It's Mostly About People

As you've seen so far, cost includes the direct and indirect resource expenditure (predominately the human element) required to investigate, lead, manage, and resolve an incident. It is a measurable expression of the economic footprint that a disruption leaves on the company. Again, many teams focus narrowly on speed (e.g., MTTR). Respononomics again changes the conversation by asking:

> "How much did it (or does it) cost us—in time, talent, tools, and business impact—to fix this?"

In Responsonomics terminology, we will refer to all human resources—whether technical SMEs, incident response teams, executives, support personnel, crisis communications, legal etc.—by one unifying word: labor.

> *Key Thought:* Keep in mind the three important L's of incident response cost: labor, labor, labor!

Look at the responders, their thinking, their focused conversations and their decisions as energy required to generate the forward momentum for incident resolution. This energy compels them to move through the incident with clarity and purpose, all the while keeping an eye on BI and the wider technology impact(s) of an incident. It's common for large, decentralized, enterprise level incident response teams to struggle with seeing beyond their individual silo in terms of incident impact. The million dollar question then, quite literally, is how much energy cost is required in the acute and long-term sense, to consistently and efficiently respond to and mitigate high severity incidents. Let's start by putting a dollar figure on what that energy cost might look like. While this is a complex goal, it is possible to apply an imperfect model to begin to more clearly understand the labor burn rate of a pool of incident responders.

To begin the discussion of calculating cost, starting with labor, consider this baseline economic model:

$$\text{Cost} = \Sigma_{i=1}{}^{n} \text{ (Time spent by responder}_i \times \text{Hourly rate}_i)$$

Σ = sum of each responder multiplied by the time spent
□ = time spent in terms of hours contributed by each responder during the incident
n= the number of hours each responder is engaged in incident response.
Hourly rate□ = * this may be calculated at a fully loaded hourly cost (salary + benefits + equity linked compensation + overhead) to the company of a responder or responders (i), or it might just be the cash compensation to the labor resource. Simple hourly rate calculations can be made by taking an annual salary of a labor resource divided by 2080 hours the typical annual work hours for a non-exempt worker in the United States.

With metrics in general, don't let perfect be the enemy of the good. If the data points are consistent and directionally correct, trends over time will emerge. Again, this is an imperfect model, and while quite basic, it is intended as a starting point to build a model based on the specifics of your company. We will be revisiting and expanding this model in subsequent chapters to include other elements of cost.

This model reflects a foundational labor input cost across functions—not just engineering time, but every moment of human attention, coordination, and communication that is consumed throughout the incident lifecycle. Let's look at the results of the model in a graphical representation (figure 3.2). These numbers are arbitrary but reflect the concept of determining total labor cost for an incident. The figure depicts only the cash compensation costs for a group of labor resources found on the right side of the graph (not fully loaded salary + benefits + equity linked compensation + overhead). You can more accurately assign the costs from your company based on specific compensation data if you choose.

As you see on the left side of the image, each bracket of labor resource type has an hourly rate assigned for work performed during a normal workday. The right side of the graphic shows the same group of available labor resources, now configured as an incident response team for a small incident. You can see that we have calculated an hourly burn rate for that group when called upon to respond, shown to be $502.00/hour. If you expand or shrink the labor pool, or change the types of responders that are dispatched, you can see the potential impact to the hourly labor burn rate. You can also begin to see the impact of having too many lurkers hanging around, or higher priced employees engaged that may not be key contributors. Again, we are not advocating for more or different headcount, we are making the point that it is possible to identify the cost of the numbers and type of labor resources that respond to your incidents.

Example of Variable Labor Cost for Incident Response

Normal Operations			Incident Response Team		
Title	Hourly	Salary	# of responders	Title	Hourly
Executive	$72	$150,000	2	Executive	$144
Director	$57	$120,000	1	Director	$57
Manager	$48	$100,000	2	Manager	$96
Senior Engineer	$43	$90,000	3	Senior Engineer	$129
Engineer	$38	$80,000	2	Engineer	$76
TOTAL	$258/hr			TOTAL	$502/hr

2,080 work hours per year x Hourly Rate = Annual Salary

Figure 3.2

You can play around with the numbers to see how different dispatch scenarios and rule sets govern who responds, to see what works best for the company. There is no right or wrong answer. It comes down to how much economic labor impact the company chooses to bear for a variety of incidents, both in *direct cost* and *opportunity cost* [more on this in the next chapter].

> ***Key Thought:*** Opportunity cost is a foundational principle in economics that refers to the value of the next best alternative foregone when a decision is made. In other words, what thing didn't get done because we were doing other things instead. In organizational settings, especially within operations and strategic planning, opportunity cost provides a critical lens for right sizing resource allocation.

To some degree, labor cost is linked to Severity levels. If your SEV levels are consistent and accurate, it is then possible to get an even narrower idea of anticipated cost per SEV level.

Figure 3.3 depicts the cost of our initial grouping of labor resources when laid out in different scenarios based on numbers of responders and the time committed to resolution. As you can see, our baseline number of 10 responders for 1 hour is shown at $502.00 per hour. If 10 more of the same responders arrive, it doubles the labor cost. If 20 more respond, it triples, and so on. It is imprecise for sure but illustrates the point. When there is an increase in the number of responders for whatever reason, and/or the time commitment increases, you can see the impact on labor cost per incident. It would be incorrect (and frankly irresponsible) for companies to believe that variable labor costs happen to them just because it's a high severity incident. We have heard from too many companies, prior to our engagement with them, that major incidents require "all hands-on deck." We are recommending quite the opposite. Effective incident response teams *dispatch* the right hands on the right deck at the right time, based on what the incident is presenting.

Incidents have an insidious way of highlighting organization dysfunction, such as lack of collaboration among teams, poor leadership, technical debt, organizational debt, and resource constraints. In reality, companies have the ability to control their variable labor cost— and they should—by understanding who must respond versus who wants to respond. Responsonomics creates a framework for a company's leaders to quantify variable labor costs so they can make data-driven decisions about their Incident Management program.

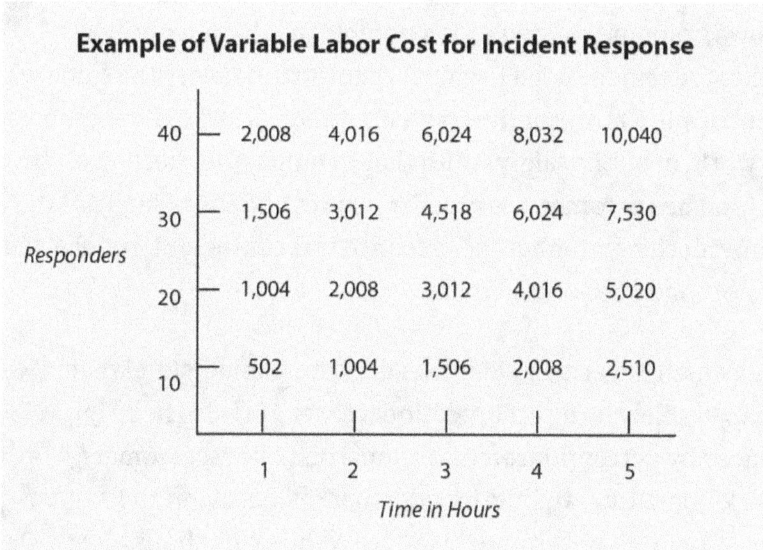

Figure 3.3

Price: What the Company Experiences and the Market Remembers

In microeconomics, price reflects what a buyer is willing to pay. In Responsonomics, price is what the company charges itself financially, and perhaps emotionally, and reputationally when a service fails—and how that financial impact and company memory shapes future behavior.

Price is often the most real and lasting part of the incident, and the sum of all the potential elements discussed previously. And as you have seen, price is not solely determined by the dollars a company spent to fix the problem. To that end, there are tangible and intangible components in the price calculation. While cost can be calculated and OC may be reasonably estimated, price is something altogether different in Responsonomics.

Price is consequence, paid in financial and reputational currency. It is shaped by perception, not precision. Some high severity incidents are so profound and far reaching that they are remembered by name, specifically the name of the company involved, such as the July 19,

2024, CrowdStrike incident that affected over 8.5 million Microsoft Windows systems worldwide. It's notable that high severity IT incidents are remembered in global IT culture in nearly the same ways that Hurricane Katrina or the Loma Prieta earthquake is remembered in the world of public safety. With that in mind, price becomes a trust indicator. The customer, the broader market and the media, not the company, decides whether the response was competent, timely, and worthy of continued engagement.

After any high-severity incident, leaders must understand that price is not shaped solely by direct operational costs. It is also heavily influenced by perception, emotion, and trust—how customers, partners, regulators, and the broader market feel about the organization in the aftermath of failure. While labor hours, lost revenue, and recovery expenses are measurable, reputational price is driven by human judgment, which is often faster, harsher, and longer-lasting than financial accounting.

The following questions provide guidance for evaluating not only what occurred internally, but how the incident was experienced externally:

- Was the response timely, transparent, and respectful? Did the company demonstrate urgency and accountability in both internal operations and external communications, or did stakeholders feel ignored, misled, or minimized?
- Did the company project competence and control, or confusion and reactivity? Markets reward organizations that appear disciplined under pressure and punish those that appear overwhelmed, disorganized, or unprepared.
- Was the outage emotionally disruptive for customers and partners? Did it prevent them from conducting business, meeting obligations, serving their own customers, or protecting their reputations? The more personal and visible the disruption, the greater the emotional residue.
- Did the incident become part of the public narrative? Was it

amplified through social media, industry forums, or mainstream news outlets, transforming a technical failure into a reputational event?

Emotional responses shape economic outcomes because trust functions as a form of invisible currency. When systems fail but leadership responds decisively and transparently, trust may remain largely intact. Customers may view the incident as an unfortunate disruption rather than a reflection of organizational weakness. In these cases, price tends to remain contained within internal costs such as labor, lost productivity, and opportunity cost.

However, when the experience generates frustration, embarrassment, or fear—particularly when customers feel abandoned or misinformed—price escalates rapidly. Negative emotions convert into tangible consequences: customer churn, reduced willingness to rely on services for critical operations, increased scrutiny from regulators, higher contractual penalties, and reputational damage that influences future buying decisions. Social amplification ensures that individual frustrations can become collective judgments within hours. What begins as a service interruption can evolve into a narrative about reliability, competence, and corporate character. Once that narrative forms, organizations often spend months or years—and significant financial resources—attempting to restore confidence.

From a Responsonomics perspective, this is why incident response is not merely a technical recovery exercise but a form of emotional risk management. Speed, clarity, empathy, and visible leadership are not soft skills; they are economic controls that directly influence how high the ultimate price tag of an incident becomes. Effective response, then, protects not only systems, but perception. Poor response multiplies dollar costs far beyond the outage itself.

Consider this hypothetical incident to reinforce the point: A regional financial services platform experiences a multi-hour service outage during peak business operations. Transaction processing halts, customer dashboards fail to load, and support queues immediately spike.

Internally, technical teams work aggressively to restore systems, but externally, communication is slow and minimal. Customers receive a brief status message stating that "engineers are investigating," with no estimated time to resolution or acknowledgment of business impact. As the outage stretches into hours, frustration grows. Small business clients begin posting screenshots of failed transactions on social media, expressing concern about missed payroll deadlines and delayed payments to vendors. Industry forums speculate about data integrity and security implications. Local news outlets pick up the story, framing it as a reliability failure during a critical business window.
When service is finally restored, no formal explanation is offered—only a generic message confirming resolution.

In the weeks that follow, customer churn increases. Several enterprise clients invoke contractual service credits. Prospective customers cite the outage when choosing competitors. Although the direct operational costs of the incident were limited to recovery labor and lost transaction fees, the emotional fallout produced far greater economic consequences through lost trust and reputational damage.

The technical failure was temporary. The market's memory of it was not.

This is why price in Responsonomics can be described as a subjective multiplier of cost as well as a specific dollar figure. The same incident, resolved in two different ways, can result in dramatically different outcomes.

There are really only two fundamental ways a company chooses to experience the inevitable price of any incident response:

- Upfront, through preparedness, responder training, coordination, and a solid company culture that supports and invests in Incident Management.
- Or afterward, in damage control, apology credits, customer loss, reputational decay, and erosion of investor confidence.

Even when labor and OC are tightly controlled, a poorly managed and/or communicated incident can impose a price that outweighs all other financial models.

> *Key Thought:* Cost is what you pay. Price is consequence paid in reputational and financial currency.

Designing for Lower Price

Responsonomics asserts that price can be intentionally designed—not just endured. Companies must treat price as a variable they influence, not a fate they absorb, and adopt the perspective that every IT incident is just a business problem in disguise.

Leadership teams should forecast price the way they forecast company revenue or expenses, by modeling reputational exposure, customer impact, and contractual obligations. Central to this premise is a clear realization and acceptance that high severity incidents will occur. Simply put, if a company has operations, it will, by nature, also have incident operations. It goes with the territory.

Price connects incident response to business risk. The next chapter will explore how these financial, operational, and reputational impacts accumulate—and how leaders must model them to make smart investments in resilience.

4 MODELLING OPPORTUNITY COST

When you learn to measure the work that never happened, you begin to see the true financial impact of every incident—not just in dollars spent, but in value destroyed

A comprehensive incident cost analysis should include all direct labor costs as well as an acknowledgement of the OC to business-related productivity. According to Becker (1965), OC is not merely a theoretical construct but a quantifiable metric that captures the trade-offs inherent in every business choice—from labor hours to capital deployment.

> *Key Thought:* While difficult to accurately quantify, opportunity cost is most certainly as legitimate as any other easily measured cost.

Here is a basic example: If an engineer assigned to a product team spends 100 hours a year on incident response activities that could have been prevented through optimized observability, or some other reduction on the supply side of incident response, that 100 hours is lost productivity for product development, product improvement, new features, etc. A Responsonomics driven approach to incident response aims to keep the people doing what the company hired them to do in the first place, which is to complete specific tasks and projects that increase the value of the business. Every employer expects some multiplier of value creation to the business above and beyond the cost of the employee. In other words, if a company hires an engineer for a product team at $100,000 per year, the implied expectation is that the engineer generates at least that much value for the company, and ideally much more in terms of new products, new features or better user experience. In fact, with Responsonomics, you can determine the average annual revenue per employee and match it against the annual cost of the employee. Across most organizations, particularly in the technical teams, companies generate annual revenue of 1-5x the annual cost of the employee. Scholars such as Buchanan and Tullock (1962) have emphasized that understanding OC improves marginal decision-making, as it prompts leaders to consider both explicit expenditures and implicit alternatives.

In high-stakes environments like incident response—where time, attention and resources are finite, OC becomes particularly salient. In fact, OC, if not acknowledged, stays silent in labor cost allocation. Ultimately, organizations that consistently assess decisions through the lens of OC are better positioned to align response actions with strategic priorities, maximize utility, and minimize hidden inefficiencies.

> *Key Thought:* Incident responses that interrupt normal work should be worthy of the time investment.

OC may be hard to measure precisely but easy to feel. Over time, repeated incident-driven context switching erodes normal work progress and contributes to burnout.

Incidents don't just cost money—they redirect attention and human involvement, and attention is a limited and valuable resource. The key Responsonomics message here is that spending time responding to things that aren't true incidents take precious work time.

With OC in mind, we can expand the basic labor cost model to include a value for OC. Again, this may be difficult to quantify, but valuable to consider. If your company has a lot of lurkers (bodies) that respond, but only a handful of people truly contributing to the resolution (players), you have a poor body to player ratio. This is a valuable place in the incident lifecycle for *specific marginal improvements*—ensuring that you're dispatching the right responders for the problem you are trying to solve.

> *Key Thought:* Specific Marginal Improvements are small positive changes to the overall process of incident management and response. Deliberate and connected process improvements all along the continuum of response will have a significant positive overall impact in incident resolution. Take every opportunity to make specific marginal improvements, as small changes made every day add up to significant improvements over a long period of time. Don't try to eat the elephant of 'getting better' in one bite—break it up into many smaller bites!

There is no magic number to plug in as OC and quantifying it may be different for each company. However, allowing any employee to join an incident Slack channel or adding links to invitations to join incident video conferencing channels like Microsoft Teams or Zoom, encourages lurking in the shadows. Invitations to join a response should only be sent to those that are truly necessary for the resolution. Others can certainly be notified of an incident but perhaps not invited to join for resolution. The fact is that more is not always better when it comes to incident response. The passive observers seldom contribute to incident resolution and often are there because they are not getting good updates and actionable information through regular channels. Effective and timely outbound communications from the incident are kryptonite to lurkers, and one effort to help reduce OC.

Yes, calculating OC may be subjective and yes, it may not be 100% accurate and yes, it isn't perfect, but you can adapt it however you like. It is illustrative to input some value to OC (OC in the formula below) to acknowledge its presence to better understand demand costs. Our example formula then expands to this:

$$\text{Cost} = \Sigma_{i=1}{}^{n} (\text{Time spent by responder}_i \times \text{Hourly rate}_i) + \text{OC}$$

First, we will look at OC in this model as a placeholder for a narrative that provides context around an incident rather than a dollar figure. Next, we will demonstrate the concept numerically. You can choose which approach fits your company culture. In some cases, a narrative expression of OC may provide more meaningful context around an incident and set the table for process improvement initiatives.

> *Key Thought:* It is not realistic to believe that incident response time commitments will ever be zero. IT service disruptions and outages will occur and there will be legitimate needs for technical talent to resolve them. The idea here is simply to understand the labor costs to a certain degree so that changes and improvements can be measured over time to see how those changes impact the largest line item of incident response—labor.

Here is a scenario depicting how OC can be identified and folded into the basic financial impact model with a descriptive narrative. Our hypothetical company, LuminaTech, is just starting to adopt a Responsonomic philosophy, hoping to find specific marginal improvement opportunities as a result of historically inefficient response to high severity incidents.

LuminaTech is a mid-sized software company known for its sleek data visualization tools. At the time of the incident, the product engineering team had been working for months on a long-awaited feature—real-time dashboard syncing. It was the kind of release that could give them a competitive edge, and customers were excited.

Then came the outage.

On Monday at 08:10 am Pacific Standard Time, a cascading failure in one of the cloud data ingestion pipelines triggered a series of downstream alerts. Within twenty minutes, a SEV-1 major incident was declared [Incident #23690] and multiple Slack channels filled up. 12 Engineers from five different teams jumped in, along with a few director level people and three senior leaders, mostly lurking to stay informed. Debugging logs flew by, screen shares started, and two different Zoom calls were initiated. Coordination was poor, and roles and responsibilities weren't clearly assigned. Instead of a tight, efficient response, it became an open forum of guesses, opinions, and dead ends with no clear Incident Command. There were no predictable outbound communications to key internal stakeholders.

The incident was eventually resolved, eight hours later. Many of the engineers working on the new release were pulled into the incident for the full eight hours. They then spent much of the next two days on the postmortem and root cause meetings.

The release planned for Friday quietly slipped into next week... and then the next sprint... and then the backlog.

No one ever filed a ticket for the time lost.
No one ever invoiced the delay.
No budget line item ever reflected the cost of lost momentum.

But everyone felt it:

- A key customer who had been promised the feature emailed to say they were "tired of waiting and now forced to evaluate alternatives."
- The sales team, unsure of the new release date, pulled the feature from their demo deck, and had to figure out a strategy to handle questions for a webinar scheduled for Wednesday.
- On the engineering team, burned out from firefighting and the finger pointing that ensued, morale dipped. Progress slowed. Frustration simmered.

A sample OC narrative for this incident may look something like this:

"For Incident # 23690, declared at 8:30am Pacific Standard Time,

> the full platform engineering team, including twelve principal engineers, was diverted from pre-scheduled architecture reviews and testing for a new customer onboarding pipeline projected to accelerate top-line growth by $3.5M over the next two quarters. The twelve engineers spent eight hours each in active incident resolution and approximately twelve hours each of time assembling the necessary information for a postmortem. Prelaunch meetings were postponed to accommodate these activities, resulting delayed testing and final build, causing the build cycle to shift by at least one week—introducing risk to the onboarding timeline for two strategic clients in negotiation."

In this case, the OC narrative sets the table for a much sharper dive into the impact of the response against the adverse impact to the projected top line revenue growth. If the labor cost was modelled along with this, and discussion occurred with sales to really understand the customer pain and hesitation, along with the impact to the engineers time and well being, it would be possible to see from a high level that more was at stake than just fixing the problem. Your use cases may be different but taking a moment to start thinking beyond the obvious parts of a resolution effort begins to elevate an understanding of what opportunity costs really are.

As an incident response leader, this presents a golden opportunity to make operational maturity based improvements while the pain of the incident is still fresh in the minds of senior leadership. To close the loop on the identified OC issues, the AAR could also include process improvement recommendations that may look something like this:

Finding #1: Need to implement a Precision-Based Dispatch Framework

Problem Addressed: This incident illustrates the economic impact of the over-engagement of high-leverage talent in incidents that do not require their time or expertise.

Recommendation: Adopt a role-based dispatch matrix for SEV1 and SEV2 incidents that aligns specific types of issues with pre-designated first responders and only escalate to specialized engineers after structured triage. Leverage a lightweight triage coordinator role to filter incidents before wide-scale paging occurs.

Responsonomics Alignment: This minimizes labor cost and protects OC by keeping strategic contributors focused on roadmap execution. It also improves the body-to-player ratio, a core Responsonomics metric.

Finding #2: Need to introduce an "Opportunity Cost Narrative" in Postmortems

Problem Addressed: OC remains an invisible but highly consequential economic loss. This has not been a focus in the traditional postmortem and AAR activities.

Recommendation: Expand AAR templates to include a section titled OC where team leads briefly describe what workstreams were paused, delayed, or deprioritized due to the incident. This section should focus on strategic initiatives or customer-facing outcomes.

Responsonomics Alignment: Codifies OC into the financial impact model—not as a number, but as a narrative—thereby raising executive visibility and supporting smarter resource trade-off decisions.

Finding #3: Improvement Needed in Outbound Communication to Key Stakeholders

Problem Addressed: Incident channels draw excessive passive observers due to poor outbound updates, increasing labor and opportunity cost and introducing potential distraction to the resolution effort.

Recommendation: Create a communications lead role during high-severity incidents tasked solely with timely, high quality, outbound updates to stakeholders every 15–30 minutes.

Responsonomics Alignment: Potentially reduces the number of responders and supports better economic containment by shrinking the incident's labor footprint.

As you can see, including OC is valuable to highlight the silent impact of having too many people, spending too much time, with no real *return on investment* (ROI) of their involvement.

Here is an example of how a more specific numerical value may be implemented as OC. This is for illustration purposes, and the main point is to focus on the process and logic rather than the numbers. In the LuminaTech example, we see that there is also a pathway to include OC as an assigned percentage value:

In this case, we will evaluate the direct cost for the 12 key responders × 4 hours × $125/hour = $6,000 (labor Cost) and estimate the OC for the others that attended. The OC is estimated at 60% of base labor cost for the 12 responders = $3,600. The total then, including OC is $9,600. The selected 60% OC factor is arbitrary in this case, and you may choose to use other numbers. The key thing is consistency so that over time, trends will become clear. At the end of the day, the total cost of $9,600 reflects not only the labor spend, but a tangible acknowledgement of loss of strategic momentum. LuminaTech's key engineering resources were pulled from core business tasks, delaying revenue-generating features and weakening product velocity. When incidents disrupt focus, they quietly erode innovation capacity. The Responsonomics approach ensures such hidden costs are surfaced, measured, understood and improved.

However you choose to include OC in your incident cost calculation, it should be viewed as representative even when identified as exact dollars and cents. Companies should first become comfortable with the idea of capturing OC, gain some experience, and play around with numbers, or narrative templates, and refine the approach over time in a way that resonates with the company culture.

With all this talk about labor, cost, and protecting the reputation of the company, you may be thinking: "Is there ever a time to knowingly over-invest in resolving a high severity incident?"

The short answer is yes, but only when it is intentional and there is a high need to preserve the customer trust equation or when some other mission critical piece of the business is threatened. In Responsonomics, cost management is not solely about minimizing expenses—it is about optimizing the economic outcome of every incident. While efficient response remains a core objective, there are situations where a deliberately expensive response is the most rational and strategic decision. This is because things like customer trust function as a form of economic capital: It directly influences future revenue, customer retention, and market reputation.

In high-severity incidents, the customer trust equation can outweigh traditional cost controls. If trust is eroded, downstream effects such as customer churn, lost renewals, elongated sales cycles and negative market perception can dwarf the immediate labor or tooling costs of an escalated response. Within Responsonomics, you can apply this concept as the *Trust Erosion Price (TEP).* The notion of a trust equation was first introduced in 2000 by David Maister in his book The Trusted Advisor. When the TEP—the quantifiable and reputational consequences of delayed or inadequate response—exceeds cost of an expensive response and resolution, paying more is not wasteful. Quite the opposite, it is protective and warranted and organizations are effectively purchasing preserved trust and future revenue stability. Responsonomics and incident response intersect when it falls to the Incident Commander or *Unified Command* to make technology resolution decisions with a strong understanding of the threat to the business as a whole. While this book does not specifically address Unified Command, you can find more information on the topic in our

Incident Response textbook *Incident Management for IT Operations*, (O'Reilly Media, 2017). As a client of ours once said, "Yes, we'll throw everything we have at this problem today because it's cheaper than losing the trust and risking the churn."

In some cases, a company may choose to break out TEP costs separately from the basic financial impact model and ultimate price calculations to determine not what was lost, but what was saved. This metric is used by some fire departments to determine the value of fire suppression. Here is an example: If a local fire department responded to a working fire within five minutes, aggressively attacked the fire and stopped the spread into other areas of a multi-million dollar structure, the price of extinguishment could be arrived at by assigning the traditional labor, materials and time cost during the active firefight, but would be much more compelling to also include the value of what wasn't lost (saved) due to an efficient response with adequate staffing.

The critical shift lies in making these decisions consciously, backed by data, rather than allowing habitual over-response to erode efficiency. This is also an approach to take when compiling a business case for increased or continued financial, political, or cultural support from the company.

Yes, these incidents cost x dollars but here is the protected value of trust for high value customers or other mission critical business functions. It changes perspective from incident response being viewed as a cost center versus a strategic defense of the company's value and financial stability.

> *Key Thought:* Cost is not a side effect of incidents—it is the central economic signal that tells us whether our teams are efficient and our practices are aligned with business value. By modeling cost transparently and using it to drive decisions, teams move from reacting to incidents to designing more cost-effective, resilient systems—the true goal of Responsonomics.

Branding the incident response team as "value defenders" rarely occurs in companies, but it is exactly what they are doing when the response is strategic and intentional!

MTTR: Why Time Alone Doesn't Tell the Whole Story

MTTR has long served as a primary performance indicator in IT incident response. At first glance, it appears to provide a clean, quantifiable measure of operational efficiency based on this logic— the shorter the MTTR, the faster the fix. The faster the fix the better the response and the less adverse the BI. But there is much more to this story! Within the framework of Responsonomics, MTTR can be viewed as incomplete—it tells us how long something took but says nothing about the real cost or impact.

Consider the following example in Figure 4.1, where two incidents share the same MTTR of two hours:

Metric	Incident A	Incident B
Responders	1 engineer	8 engineers 2 managers
Total Time	2 hours	2 hours
Total Cost	~$200	$4,000+

Figure 4.1

At face value in this example, MTTR implies parity. Both incidents were resolved— presumably efficiently—in two hours. But in economic terms, they are profoundly different. Incident B consumed 10x the labor cost, introduced more overhead cost, and likely caused more downstream disruption across teams, creating OC impacts. Again, what things could the eight engineers and two managers been doing instead of this? Perhaps they were needed to resolve this, which justifies the labor cost, but only the company and the situation can make that determination. This example illustrates a fundamental flaw in over-relying on MTTR: it is volume blind. It does not account for the intensity of labor input, the level of coordination complexity, or the OC incurred by pulling multiple high-value resources off their primary responsibilities. Time still matters of course, but it must be paired with economic density: the concentration of cost, skill, and attention spent during that time. In other words, two hours at $200 is clearly not the same as two hours at $4,000—and your metrics should reflect that difference. While operational metrics like MTTR may provide some value, they are not stated in the language of business. In other words, viewed from a holistic economic lens, submitted to business leaders in the same terms and approach the guides the business.

5 PRICE AND BUSINESS IMPACT

Every incident creates two stories—the technical fix and the business impact. Ignore the second, and you'll never understand the real price of fixing the issue

In Responsonomics, we distinguish between two distinct but interconnected concepts: cost and price. While cost reflects the internal resources consumed during an incident, price includes other tangible and intangible factors. As you've seen, price may be expressed quantitatively (e.g., lost revenue, SLA penalties, refund requests, credits, regulatory fines) or qualitatively (e.g., customer frustration, trust erosion, reputational damage, elongated sales to close cycles). By understanding and influencing how incident response intersects with customer perception, contractual obligations, and business impact, leaders can prevent some level of financial and reputational damage before it occurs.

> *Key Thought:* Every IT incident is just a business problem in disguise.

Every technical decision made during an incident—where reliability is invested, how quickly teams mobilize, what is communicated, and how impact is evaluated—either contains or amplifies the ultimate business consequence. In Responsonomics, price is therefore not an abstract outcome but a variable leaders can actively influence through disciplined operational choices. The following five tactical practices translate this economic principle into concrete response behaviors designed to reduce reputational exposure, preserve trust, and prevent unnecessary financial loss.

The following five tactical practices translate price control into an operational strategy of loss prevention.

Tactic 1: Prioritize Customer-Facing Systems in Reliability Investments

→ Protect the systems where failure is instantly visible and emotionally amplified.

Not all systems carry equal economic risk. Failures that directly affect customer access, payments, onboarding, or core service functionality create immediate frustration and public exposure. In contrast, many backend degradations remain invisible to end users and therefore generate far lower reputational impact.

Customer-facing outages tend to produce disproportionate price because they involve:

- Real-time visibility of failure
- Immediate disruption of revenue or service delivery
- Rapid amplification through social media and support channels

From a Responsonomics perspective, this creates an asymmetry of risk: a small technical failure can trigger large economic consequences when it touches customer experience pathways.

Responsonomic Guidance:
Anchor reliability investments—such as redundancy, monitoring, load testing, and failover—around systems with the highest external exposure. Dependency graphs matter, but customer visibility matters more. Even a single impacted enterprise customer may represent thousands of downstream users and significant reputational risk.

Tactic 2: Ensure Responders Understand SLA and Uptime Obligations

→ Clarity in performance commitments prevents accidental financial and trust erosion.

SLAs and SLOs are not merely technical benchmarks; they are financial contracts that translate outages into tangible penalties. Ambiguity in how uptime is defined, measured, or communicated increases the risk of disputes, refunds, and damaged relationships.

When expectations are unclear, even well-intentioned maintenance or partial outages can be perceived as breaches of trust.

In high-severity incidents, internal response timing metrics alone are insufficient. Urgency should not be dictated by contractual minimums but by reputational risk. Just as emergency services do not wait the full allowable response time when danger is immediate, incident response must operate in real-time urgency mode when trust is on the line.

Responsonomic Guidance:
Clearly define uptime calculations, performance thresholds, and penalty conditions. Track them continuously and ensure responders and leadership understand their financial implications. The cost of clarity is consistently lower than the cost of misinterpretation.

Tactic 3: Implement Proactive, Transparent Communications During High-Impact Events

→ Trust is shaped as much by messaging as by technical restoration.

Customer sentiment during incidents is driven not only by system availability but by how openly and confidently organizations communicate. Silence, vagueness, or defensiveness increases emotional frustration and narrative speculation. Clear, timely updates reinforce competence and accountability—even when services remain impaired.

Effective incident communication reframes disruption from perceived chaos to managed recovery.

There is not inherently “good” or “bad” news in incident response—only the reality of progress. Consistent updates reduce uncertainty, calm stakeholders, and prevent misinformation from filling the communication vacuum.

Responsonomic Guidance:
Treat communication as a core response function, not a burden or interruption to the resolution effort. Develop process for customer-facing updates alongside technical resolution plans. Transparency, when paired with demonstrated competence, is a powerful economic control.

Tactic 4: Integrate Customer Impact Forecasting into Post-Incident Reviews

→ Measure not just what broke, but what the disruption cost in trust and business outcomes.

Traditional postmortems focus on root cause and remediation. While necessary, this approach overlooks the broader economic footprint of incidents. Without assessing customer experience and market impact, organizations underinvest in the areas that truly drive price escalation.

A Responsonomics-aligned review expands analysis to include:

- Estimated lost revenue
- Support volume surges
- Churn indicators
- Social and public sentiment
- Honest evaluation of response effectiveness

This transforms postmortems from technical reports into strategic learning tools.

Responsonomic Guidance:
Add a dedicated customer and market impact section to every high-severity incident review. Quantifying emotional and reputational

effects informs smarter reliability investments and justifies proactive prevention efforts.

Tactic 5: Use *Tabletop Exercises* to Simulate Reputational Risk Scenarios

→ Prepare the entire organization for high-visibility, high-consequence incidents.

Most technical teams rehearse system failures. Few organizations rehearse the cross-functional chaos that accompanies public incidents—where legal, communications, executives, customer support, and regulators must coordinate in real time.

High-severity events rarely remain confined to IT operations.

Effective simulations incorporate:

- Executive decision-making under uncertainty
- Regulatory and disclosure obligations
- Media and customer communications
- Internal information control
- Support response surges

These exercises reveal bottlenecks, misalignments, and emotional escalation risks before real incidents occur.

Responsonomic Guidance:
Run regular tabletop exercises focused on reputational and regulatory exposure, not just technical failure. Test communication channels, authority flows, and cross-department coordination under realistic pressure.

One of the core insights of Responsonomics is that the BI of an incident often dramatically exceeds its tangible internal cost. This asymmetry is especially pronounced in customer-facing systems, country-specific regulatory environments, and high-visibility industries such as healthcare, energy, gaming or finance.

Price calculations, then, would be incomplete without including some formal acknowledgement of BI in our financial model. Like OC, BI may be subjective—and in some cases more of an educated guess but does help a company acknowledge the often omitted internal impacts of an IT failure. Think of a BI summary as an investigatory action that steps outside the borders of an incidents and takes a 30,000' view of the incident and the tentacles it may have into any and all areas of the business. Ironically, companies often have significant discussions about BI during the response, but it typically centers on how the issue is impacting a customer base. When looked at this way, BI appears to be a consequence of the incident, not an integral part of it While customer impact is important, it is equally important to acknowledge the impact the incident is having on the operations of the company in terms of labor commitment, leadership clarity, executive involvement, etc. To that end, the baseline Responsonomics economic impact model can be revised like this:

$$\text{Cost} = \Sigma_{i=1}^{n} (\text{Time spent by responder}_i \times \text{Hourly rate}_i) + \text{OC} + \text{BI} = \text{Price}$$

Here is a scenario to illustrate the point and provide you with a sample approach to expressing BI in more detail:

Payment Processing Outage at Fintrax Solutions

Company Profile

- Industry: Financial Technology (B2B Payments Platform)
- Size: 2,100 employees (300 engineers, 35 incident responders)
- Annual Revenue: $450M
- Customer Base: 9,000 enterprise clients, primarily in retail and logistics
- Primary Service: Real-time payment authorization and settlement services

The Incident

On a Monday morning—one of the busiest payment windows of the week—Fintrax Solutions experienced a failure in its distributed database cluster supporting transaction processing. The misconfiguration stemmed from a weekend patch rollout that introduced a cascading indexing error.

Detection lagged. The alert fired immediately, but poor observability meant noise drowned the true signal. By the time a clear SEV-1 was declared, thousands of transactions had already failed. Assembly swelled to fourteen engineers, including four senior architects pulled from ongoing strategic projects. The technical resolution—rolling back the patch and resyncing the affected cluster—took less than an hour, but total MTTR stretched to five and one half hours because of fragmented troubleshooting, poor escalation paths, and duplicated efforts.

From a purely technical perspective, the incident was resolved before the day was out. But from an economic and business perspective, the fallout was severe.

BI Summary for Executive Review

When evaluated through the Responsonomics framework, the true price of the incident was not simply the labor costs. It was the broader financial, reputational, and strategic risks that cascaded through the business. Here's the tale of what really happened:

Direct Labor Cost

- 14 responders × 5.5 hours × $100/hour = $7,700

Opportunity Cost

- Four senior architects diverted from a $2M strategic feature launch → feature release delayed two weeks, deferring ~$250,000 in projected revenue.

- Customer success managers pulled off new client onboarding → two enterprise clients ($120,000 ARR combined) delayed integration by 30–60 days.

Business Impact

- Customer Trust: 17 enterprise clients experienced failed transactions, resulting in escalations to executive sponsors. Two high-value accounts expressed doubts about contract renewal at year-end.
- Reputation: Negative chatter appeared on LinkedIn and industry Slack channels, raising concerns among prospects about platform stability.
- Strategic Risk: A major retail customer paused discussions on expanding usage, citing lack of confidence in reliability—a potential $1M ARR expansion opportunity now at risk.
- Cultural Cost: Burnout indicators rose; three engineers reported intent to leave if on-call stress persisted.

From a senior leadership perspective, this incident demonstrates why technology failures cannot be measured in isolation. The total modeled economic price of the incident is not $7,700—it exceeds $1.3M in quantifiable financial exposure, plus long-term reputational damage that could impact renewal cycles and market positioning.

Following is a sample executive level BI report for this incident. You can use this as a template to create your own BI worksheet for incident cost evaluation.

Incident Number: FIN-SEV1-2023-0417
Date/Time Incident Occurred: Monday, 8:45 AM – 2:15 PM EST
Severity Level Assigned: SEV-1
Number of Responders Engaged: 14 (including 4 senior architects)

Incident Narrative Summary

On a high-volume Monday morning, a database cluster misconfiguration caused cascading payment processing failures across the Fintrax platform. Poor observability delayed accurate detection, while unclear escalation paths led to duplicated troubleshooting and inflated responder headcount. Although the technical fix took under one hour, the full incident lasted 5.5 hours. The economic fallout extended far beyond labor costs. It resulted in delayed revenue delivery, deferred client onboarding, reputational damage.

Business Impact Dimensions

Revenue Impact:

• $250,000 deferred revenue from two week delayed roadmap feature release and $120,000 ARR delayed from two enterprise clients whose onboarding was stalled.

Customer Trust Impact:

• Seventeen enterprise clients experienced failed transactions, leading to escalations.
• Two high-value accounts expressed doubt about contract renewal at year-end.

Reputational Impact:

• Negative chatter appeared on LinkedIn and industry Slack communities.
• Sales team reported prospects citing "stability concerns" in pipeline conversations.

Strategic Impact:

• $1M ARR expansion deal with major retail client paused due to instability concerns.
• Competitor highlighted incident in sales engagements, creating market risk.

Cultural Impact:

- Three engineers expressed intent to leave citing unsustainable on-call load.
- Team morale negatively impacted by long hours and duplicated troubleshooting.

6 OBSERVABILITY AS AN ECONOMIC LEVER

Better signals mean faster triage, lower costs, and preserved trust

When thinking about BI, it's important to note that the impact of an incident begins well before the first responder joins the call. In fact, the moment that an alert is detected, the incident already has an operational head start on the incident response team and a BI headstart on the company. Every second of disruption, even before you know about it, quietly starts the cash register tabulation of direct and *indirect costs*. Poor detection and inaccurate alerts are among the most expensive—and most preventable—drivers of incident cost. They are also the most common issues we encounter during customer engagements. We've heard many times that monitoring thresholds are different for different teams, and/or they were set by individuals or teams that are no longer at the company, and/or are arbitrary and not applicable to the current technology environment. The more complete way to look at observability is as an important step in strategic cost-control. By slashing time-to-detection, eliminating unnecessary dispatch call outs, and enabling smarter triage, observability has a direct impact on the early economics of incident response.

> ***Key Thought:*** Observability is the first tangible intersection between incident management and incident response.

Observability is the inflection point where controlling supply can directly impact cost in the incident response marketplace. In some ways, observability is a bit like fire prevention—difficult to quantify what didn't happen, but certainly valuable to the overall process of fine tuning what truly should enter the incident response process.

This chapter explores at a high level the economic impact of observability, explains why reliable and stronger signals reduce incident cost, and outlines how teams can sharpen detection to minimize time and labor cost.

Observability vs. Monitoring: Beyond System Health to Economic Intelligence

In IT operations, incidents are discovered either by people or machines. From a technology point of view, monitoring and observability are often treated as interchangeable. However, within Responsonomics, the distinction is not just technical—it is economic. The way an organization invests in detection determines whether it reacts to failures or strategically responds to disruption. Reacting, in incident response terms, is a rapid, sometimes emotional, and less than predictable way to approach a situation. For example, a person whose house is on fire may react with high emotion and perhaps take an irrational action of trying to extinguish the fire with an inadequate amount of water from a garden hose. It's clear there is more fire than water, but the person is doing something, even though it isn't ultimately productive. Responding on the other hand, implies a more structured, direct and intentional approach. For example, the fire department is dispatched to the fire, establishes command, and directs incoming resources to specific tasks to put out the fire. Responding is rooted in the acceptance that incidents can and will occur, and the company's philosophy is structured accordingly. Again, this first step in the incident lifecycle is vital to get right as it has a direct influence on every other thing that happens afterward. Operationally, the difference between monitoring and observability can be broken down in Figure 6.1.

Approach	Core Question	Example
Monitoring	"Is this thing broken?"	CPU > 90% triggers a static threshold alert
Observability	"Why is this behaving this way?"	Correlates metrics, traces, logs, and events to diagnose causality

Figure 6.1

As you likely already know, observability is much more comprehensive and far reaching from a technology point of view, with a goal of providing contextual intelligence via the monitoring tools. Ideally good observability allows IT and business leaders to align response priorities with BI instead of reacting blindly or simply based on an emotional reaction to incident resolution. A mature observability practice transforms raw telemetry into decision intelligence in the following ways:

- Helps responders focus on the highest-value problems first.
- Reduces unnecessary involvement of expensive senior engineers, executives and/or other parts of the company that are critical to the day to day strategic goals and objectives of the business.
- Detects weak signals before they turn into costly responses or outages.

This shift is critical for improving operational maturity. Instead of throwing all hands on all decks, organizations design response portfolios via identified severity or priority levels—choosing intentionally the types of technology failures that require a full response, where and when to over invest to protect trust, and how to distinguish between minor events and major incidents. It's certainly not without challenges, and is never 100% accurate, but observability allows the IT response team to focus on those disruptions and outages that require the full weight of the organization, whatever the reason. The result is responding to the real fires rather than false alarms.

Monitoring: A Binary Model with Hidden Costs

Monitoring has long been a foundational component of IT operations. It provides essential visibility into system status, flagging when key infrastructure components deviate from defined thresholds (e.g., CPU > 90%, disk I/O > 75%, HTTP error rate spikes). But at its core, monitoring remains a binary model. It answers the question, "Is something in the environment functioning out of its specified range?" While this simplicity makes monitoring useful for basic system awareness, it is insufficient for today's dynamic, distributed environments, where failure is often emergent, multi-causal, and context specific. Further, the specified range is set by numerous factors, but all within the control of the company. We've seen clients who set low monitoring thresholds because the purchasing process to buy more technology equipment or services is so long and arduous that the threshold kicks off a procurement process. We've seen other clients who set high monitoring thresholds so they can run their systems hot, because the CFO wants to produce high operating margins by keeping IT operating costs low. In these cases, economic principles as applied to monitoring thresholds are not uniform and are set up to mask organizational dysfunction. In Responsonomics, the limits of monitoring aren't just technical and inconsistent—they're key to the economic impact of incident response. Consider these four structural weaknesses of monitoring:

Shallow Visibility: Monitoring often detects surface-level symptoms, such as high memory usage or elevated error rates, without providing context or root cause insight. This leads to prolonged triage and may require numerous responders from numerous technical domains to piece together what happened from fragmented telemetry. Shallow monitoring contributes to high labor costs.

Static Thresholds and High False Positives: Traditional monitoring relies on hard-coded thresholds that lack context sensitivity. A CPU spike may be expected during a backup window yet still trigger an alert.

This creates unnecessary interruptions, response noise, and fatigue. It also creates complacency and perhaps instills blind spots that prevent early awareness that some technology problem is brewing but hasn't exceeded a threshold...yet.

It is Reactive, Not Strategic: Monitoring is inherently reactive. It alerts based on exceeding thresholds (however they were set) after something has gone wrong, with little insight into why it failed, what preceded it, or how to prevent recurrence. This leads to slow and inefficient assembly of responders, delayed troubleshooting and eventually, slow recovery.

Lack of Causal Correlation: In modern architectures such as microservices and Application Programming Interfaces (API) calls, monitoring fails to connect upstream symptoms with downstream effects. Engineers see alerts from multiple discrete systems but cannot trace the fault line that connects them or the overall path that creates the service. This may inhibit the ability to determine what customers may be impacted or how widespread the impact truly is.

As mentioned earlier, organizations often track the explicit costs of incidents but fail to account for the OC and BI created by poor detection. Day after day, response teams spend time diagnosing ambiguous alerts and incidents that could be resolved by one person in fifteen minutes, or have them balloon into multi-hour, multi-team efforts, inflating both direct labor spend and recovery timelines. Again, every unnecessary page or misdirected callout pulls engineering labor's focus away from the basic business roadmap.

Reactive response amplifies cost. Imagine a scenario where a degraded search experience is caused by a downstream database lock. Monitoring triggers an alert for high CPU, slow queries, and timeout errors, but fails to explain why they're occurring or how they relate. The result: three teams paged, two hours of triage, twenty+ person-hours lost, and degraded customer experience. Had comprehensive observability been in place to correlate logs, traces, and metrics, the root cause may have been identified and resolved in a much shorter time frame with much lower labor cost to the incident and OC to the company.

Poor monitoring creates and reinforces the illusion that all is well because no critical alerts have fired. This encourages passive detection postures and frantic escalation as a normalized default. A key tenet of Responsonomics is that costs don't disappear—they relocate. Monitoring-only environments shift costs downstream into response, recovery, problem management and customer service operations. Think of it this way: Monitoring may tell you that the plane lost altitude. Observability tells you whether it was turbulence, pilot error, or engine failure—and how much fuel you lost while troubleshooting.

Again, the value of observability is that it correlates metrics, traces, logs, and events to create a contextualized, cause-driven model of system behavior. Instead of answering "what is wrong," observability reveals why it's happening and how it affects the business.

Key Observability benefits from a Responsonomics lens:

- Reduces MTTD → fewer customers impacted, lower trust erosion.
- Minimizes unnecessary escalation → keeps expensive senior engineers focused on strategic work.
- Illuminates cross-service dependencies → highlighting weak areas and enabling proactive containment before incidents cascade.

The result is economic leverage. In effect, observability converts raw telemetry into economic intelligence. Industry studies reinforce this. Over half of organizations surveyed saw MTTD and MTTR improvement after adopting or refining observability, with 29% achieving 25%+ gains.

Example:
A service emits 200 alerts weekly—only 15% actionable. The result: 170+ interruptions costing an estimated $20,000–$40,000 annually per team. Low signal quality equals hidden waste.

Therefore, faulty observability erodes economic efficiency through:

- On-call fatigue and desensitization
- Unnecessary escalations
- Cognitive drain and context-switching cost

As such, observability is not a cost—it's an investment. Benefits include:

- Fewer false positives → reduced labor waste
- Faster triage → shorter MTTD/MTTR
- Empowered responders → less senior escalation
- Learning loops → prevent future failures

In a culture of shared responsibility for accurate early warning, every team contributes to cost efficient detection by:

- Defining alerts tied to business outcomes → Focus on signals that represent true customer or business risk, not just system noise.
- Maintaining and reviewing SLOs regularly → Service-level objectives must reflect realistic thresholds and customer expectations, not arbitrary uptime goals.
- Tracking and addressing alert fatigue → Teams should acknowledge the human toll of on-call fatigue, false positives, and alert actionability. Persistent alert fatigue is a cost amplifier and burnout risk.
- Participating in cross-team signal reviews → Just as teams hold code reviews, Responsonomics recommends that they must regularly audit and rationalize alerts to avoid redundancy, eliminate blind spots, and align with evolving architecture.

One of the most powerful mindset shifts in observability culture is the adoption of a test-before-page standard: Just as a company should not ship untested code, it should not ship unvalidated signals.

> *Key Thought:* Observability maturity is not just about tooling—it's about how your organization thinks about detection, accountability, availability and risk. In Responsonomics, observability is a cultural asset. It protects trust, prevents waste, and reduces the opportunity cost of every alert.

Signals should be tested for:

- Accuracy: Does it alert only when necessary?
- Clarity: Is the alert actionable by the on-call responder?
- Business relevance: Does this signal tie back to cost, trust, or opportunity?

Failure to validate signals before deployment leads to alert noise, alert fatigue for on-call resources, cognitive overload, and desensitization—all of which delay detection and degrade trust response. One way our customer base tests the link between observability and incident response is to conduct regular Mobilization Exercise (MOBEX) drills. This drill simulates an incident response in order to measure, evaluate and refine the intersection between alerting, labor assignment and the optimization of *Mean Time to Assemble (MTTA)* for the appropriate team of responders. MOBEX drills can be as simple as developing a series of simulated alerts, firing the alerts, and observing which people, teams, or business units respond and whether they meet established company SLAs.

These are diagnostic in nature and can be as complex as you want. We have found over the years that when companies take a detailed look at the relationship between observability and team assembly, there are many areas of opportunity for improvement. In a way, a MOBEX is also an observability tool, except we're observing the behavior of labor resources.

7 THE INCIDENT LIFECYCLE IN RESPONSONOMICS

All Incidents follow the same trajectory. Understanding that is critical to understanding the dependencies of incident resolution

The *incident lifecycle* provides the visible operational representation of Responsonomics and is best viewed as an interconnected marketplace of processes and human transactions. Each phase of the lifecycle represents a different kind of transaction—between people and technology, between responders and leadership, and ultimately between the company and its customers.

> *Key Thought:* The incident lifecycle is a marketplace where time, talent, and trust are traded at a cost—and every inefficiency is an economic tax on the company.

While earlier chapters went into detail on cost, price, supply, demand, OC and BI individually, this chapter unites those principles into a single visual image. As Responsonomics strives to create a predictable platform for improving operational maturity, it also sets the stage for efficient incident response. Every type of incident—from structure fires to medical emergencies to IT incidents—progress through distinct phases. By tracing these stages in sequence (see Figure 7.1), we can see how a Responsonomics approach transforms IT incident response into an efficient economic marketplace of time, labor, and trust.

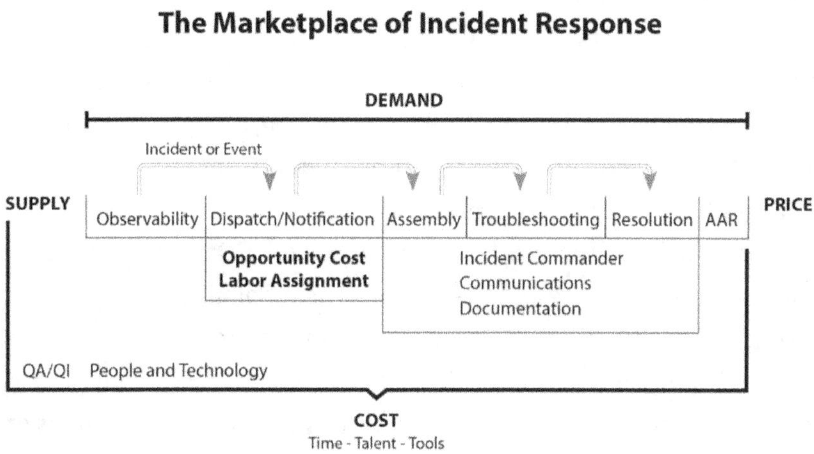

Figure 7.1

Previous chapters describe the role of each of the phases. With this graphic, hopefully, it is easy to see the relationship between supply, cost, demand, and price and where one specific marginal improvement may have impact on some other part of the incident lifecycle. As an example, every resolution activity depends first on efficient assembly, reinforcing the fact that the entire incident lifecycle is a serial process, not a parallel process. Therefore, most efforts at improving operational maturity start at optimizing supply.

The following content will serve as a linked summary of each of the phases.

Observability
Observability serves as the economic front door to every incident. A noisy alert pipeline inflates incident supply, which increases incident cost and incident price to the company. In Responsonomics, observability is not just telemetry—it is cost control. High-fidelity detection reduces MTTD, minimizes false positives, and lowers total labor expenditure. Strong observability converts raw data into actionable operational and economic intelligence.

Dispatch and Notification

Once an alert triggers, dispatch determines who responds, how quickly, and at what cost. Dispatch should be intentional and immediate. Although each incident is novel and unplanned, companies should analyze historical incident data to inform what labor resources should be dispatched to the most common incident types, resulting in a dispatch matrix for the incident response team. Certainly, not every incident type can be pre-planned, but the act of planning a dispatch matrix and identifying initial dispatching of responders is a very useful exercise. Poorly defined escalation trees or unclear SEV levels create over or under dispatching, perhaps pulling in the wrong resources or expensive resources unnecessarily. Responsonomics highlights dispatch efficiency as a labor allocation exercise: Engaging at the right skills at the right time results in the lowest sustainable cost.

Assembly

Assembly refers to the formation of the incident team. Over-assembly drives up the burn rate of labor hours, while under-assembly potentially increases MTTR and/or endangers the reputation of the company. Responsonomics frames assembly as an optimization challenge: balance the cost of additional responders against the marginal benefit of reduced incident duration. Every unnecessary participant represents an OC—time diverted from other high-value work.

Troubleshooting

Troubleshooting is the highest-intensity demand point in the incident lifecycle. This is where labor, tools, and expertise concentrate in the early stages of an incident. It is typically the most chaotic part of resolution and has a direct impact on efficient resolution and optimized MTTR. In this phase, ineffective assembly of the team, poor inter-team communication, or absent incident command multiplies cost and elongates the transition to developing a resolution plan. Frequently, disorganized troubleshooting leads to fragmentation for the remainder of the resolution effort. Said differently, technical expert resources, if not properly lead by an Incident Commander, can elongate the resolution effort, which drives up the cost and price of the incident. Responsonomics teaches us to view troubleshooting efficiency as both a technical and economic outcome.

Resolution

Resolution is the phase where cost transforms into the total price of resolution. The incident is technically closed, but the price paid in customer trust, regulatory risk, and brand reputation lingers. Responsonomics emphasizes that resolution must be measured not only in MTTR, but in labor cost and downstream financial and reputational impact.

AAR

The AAR is where economic intelligence is harvested and fed back into the company for specific marginal improvements. Beyond identifying root causes, an AAR must capture the economic story. Embedding these findings into response planning ensures that incidents feed back into systemic efficiency. In Responsonomics, conducting an AAR is not a formality—it is the profit-and-loss statements of operational resilience. It is critical to the future of the business to teach the people and the system how to continuously improve managing incidents over time.

QA/QI (Quality Assurance and Improvement)

Quality assurance closes the loop between incident resolution and future prevention. By analyzing metrics, false page rates, and escalation patterns, organizations can target the specific drivers of cost. This is the engine of continuous improvement: marginal investments in automation, training, or alert tuning compound over time, lowering total cost of incidents. QA/QI transforms metrics, data, and measurement into economic leverage.

Role Clarity as Optimization

As mentioned earlier, this is not a textbook intended to dive deeply into command and control. We do believe, however, that poor leadership and unclear roles and responsibilities during an incident significantly contribute to economic success or failure when resolving high severity incidents. To that end, we offer a few thoughts on Incident Command and the importance of a coordinated and directed resolution effort.

Some teams resist defined roles, fearing the creation of a bureaucratic culture or degrading a company's culture of creativity and collaboration. Nothing could be further from the truth. Incident response is very different than day to day business operations and requires a departure from typical corporate culture and/or the social norms that go along with it. Responsonomics reframes the structure of defined roles and responsibilities as an optimization strategy, and a clear message to the company that incident response is every bit as important as any other business unit in the company.

> *Key Thought:* Incident Command, Communications and Documentation are not just overhead expense—it is cost control. Clear roles and structured workflows reduce troubleshooting and resolution costs.

In fact, during the highest severity incidents, the rest of the business depends on the lowest price of the incident for the future of the company. To that end, clear roles accelerate decision-making, protect focus, distribute workload fairly, and support psychological safety. In short, structure and process are key features of incident response, not liabilities.

Core Roles in Incident Response

Clearly defining the goals and objectives of each key incident response role, established well before an incident occurs, avoids unnecessary complexity while ensuring coverage of critical functions.

> *Key Thought:* The goal of defined roles and responsibilities is not rigidity but precision: Allocating labor to maximize impact while minimizing overhead.

Key roles include:

Incident Commander (IC): Leads the response, owns decision-making, and delegates tasks. The IC provides structure under stress, ensuring focus, forward momentum, and reducing wasted time during the response.

SME: Provides technical diagnosis and remediation. SMEs must remain disciplined, focusing only on relevant expertise to prevent scope creep.

Communications Liaison: Manages internal and external updates (status page, leadership, customers).

Scribe (S): Captures actions and decisions for post-incident learning and accountability. Scribes, when implemented correctly, also become a potential resource for the Communications Liaison, capturing a solid base of information for the LNO to craft written or verbal briefings or messaging.

Avoiding the 'Too Many Generals' Problem

A common failure pattern in incident response is the bridge full of managers, where leadership is either overabundant, unclear, or informally contested. When multiple voices attempt to direct response simultaneously, decision latency increases, accountability blurs, and teams become hesitant to act. What appears to be increased oversight from senior level leaders not routinely engaged in the response process often produces the opposite effect: slower resolution, duplicated effort, and growing frustration among responders.

In these instances, responders must continuously assess not only the technical problem but also whose direction carries authority. Time is lost clarifying priorities, resolving conflicting instructions, and navigating interpersonal dynamics. Each pause compounds operational cost while extending customer impact and reputational exposure. Beyond inefficiency, excessive leadership presence suppresses psychological safety. When large audiences of senior leaders observe in real time, responders become less willing to think out loud, test hypotheses, or acknowledge uncertainty—behaviors essential to rapid problem-solving in complex systems. Fear of post-incident scrutiny can replace curiosity, driving overly cautious actions rather than creative resolution.

Effective Incident Management relies on a single, clearly empowered Incident Commander (IC) who owns prioritization, coordination, and decision flow. This does not eliminate executive visibility or strategic oversight; rather, it channels it through a structured command interface that protects speed and clarity. Leaders outside the IC role support the response by removing blockers, approving resources, and managing external stakeholders—not by directing technical troubleshooting or resolution.

Designating an IC early—even during low-severity incidents—builds organizational muscle memory. Over time, this reduces leadership overhead, accelerates response tempo, and preserves focus on resolution rather than governance.

From a Responsonomics perspective, command clarity is an economic control. Faster decisions reduce outage duration, minimize customer disruption, and contain reputational price. Conversely, leadership congestion inflates cost. In short, successful incident response does not require more generals—it requires one empowered commander supported by a coordinated organization.

Unclear roles also create *coordination debt*—the compounding inefficiency and error risk that grows when confusion is left unchecked. Every decision requiring three people instead of one increases cognitive load, slows throughput, and adds hidden labor cost due to slow troubleshooting or resolution plan development. Like technical debt, coordination debt is not always visible, but its impact shows up in burnout, turnover, and recurring incident inefficiencies. Typical outcomes of coordination debt include conversations going silent or conversely, having way too many concurrent discussions. Clear roles do more than streamline incident resolution—they reduce labor waste. Remember, Incident Command is not cost—it is cost control.

Remember these four key points when considering the advantages of clarity in roles and responsibilities and the value of clear leadership and coordination during a high severity incident:

- Coordination failures drive hidden costs in incident response.
- Role clarity reduces duplication, decision latency, and responder fatigue.
- Defined roles enable lean but effective response.
- Coordinated incident response is economic optimization.

8 CASE STUDY: RESPONSONOMICS IN ACTION

Responsonomics turns incidents into financial intelligence

Case studies serve as one of the most powerful tools for adult learners because they allow the reader to apply abstract principles to realistic scenarios. In Responsonomics, the value of a case study is not simply to recount the timeline of an incident, but to expose the economic dynamics hidden beneath the surface of everyday operational challenges. The purpose of this chapter is to show how the foundational concepts we have discussed so far, play out inside a hypothetical organizational setting. By analyzing the Incident Management journey of RayneCloud Inc., we will observe how recurring technical disruptions escalated into significant financial and reputational losses, and how the deliberate application of Responsonomic principles transformed chaos into cost control. Although it is a fictional company, it is not a theoretical model—it summarizes patterns that we have encountered in many of our global customers over the last decade. Readers should approach it as both a mirror (to compare against their own environment) and a template (to apply when making cost-aware, trust-preserving improvements). This scenario is certainly high level and scripted to reinforce the principles outlined in the book. Take from it what pertains to your situation and develop your own roadmap to improving operational maturity.

Case Study: RayneCloud Inc. – From Chaos to Cost Control

Company Profile

Industry: SaaS / Cloud Infrastructure
Size: 3,000 employees (400 in engineering, 40 in incident response

roles)
Annual Revenue: $300M
Customer Base: 12,000 enterprise clients worldwide
Primary Service: Multi-region computing, storage, applications and analytics serving the global financial services, financial technologies and logistics industries

Background

RayneCloud faced recurring high-severity incidents across their distributed systems. Despite rapid response times, MTTR metrics remained high and customer complaints revealed a growing dissatisfaction with RayneCloud's service. During the weekly Senior Leadership Team meeting, the new company CIO noted with frustration: "We need to do something about our incident response. I know I've only been here for a few months, but based on the incidents I've attended, we seem to flail when it comes to high severity situations. Customer satisfaction is taking a beating and I'm getting an earful from the response team. In talks with our sales team, I also learned that we recently missed onboarding a new customer because they perceive our service as unreliable. Another is slow rolling their integration due to our last SEV-1. I recently watched a webinar on Responsonomics, and I think we should deep dive into how we get better at this!" She immediately commissioned an internal audit and process review of the entire incident response process, and the initial report found the following:

- Over 40% of SEV-1 incidents had approximately 12 or more responders. An informal poll of the Incident Commanders for those incidents reported that only two or three responders were actual key problem-solving contributors. The rest were there out of curiosity and to get information because there was no structured process for updating internal communications to the executive team.

- SMEs were routinely pulled off roadmap features, delaying testing and deployment.
- RayneCloud experienced slow assembly of responders due to inconsistent staffing levels and not enforcing company establish SLAs for response.
- RayneCloud didn't know which customers were provisioned on which systems.
- No modeling existed for determining the actual labor cost of incident response.
- One incident resulted in a missed onboarding for two $50K annual clients = $100,000.
- Another incident resulted in a postponed data partnership integration (risking market advantage) = $50,000 (conservatively).

The CIO met with various directors, product teams, and customer facing communications teams and determined the following course of action:

1. Complete a labor impact study of high severity incidents that occurred during the prior calendar year.

RayneCloud completed a baseline labor cost analysis using the following data:

- Typical incident is attended by an average of 12 responders.
- Average loaded rate per responder is $125/hr.
- Average MTTR: 4.2 hours.

Based on that data, RayneCloud leadership modeled labor costs this way:

Labor cost/incident=12 responders × 4 hours × $125/hour=$6,000/incident

$6,000 per incident ×40 incidents/quarter ×4 quarters/year =

$960,000/year in SEV-1 incident labor spend.

2. The leadership team began an initiative to identify OC.
   After much research and many robust and honest discussions, it was learned that OC included the following:

   - Senior engineers frequently missed planned product release sprints and meetings due to incidents and compiling information for customer requested AAR's.
   - Customer success managers were diverted from onboarding 2 new high-value clients due to reputational damage incurred during a series of outages.
   - In one case, it was clear that there was revenue lost from a delayed go-live product launch: estimated $75,000.

3. The leadership team assessed the cost of poor coordination and lack of clearly defined Incident Command.

   - IC not assigned → estimated decision latency of >30 minutes in most incidents, resulting in elongated MTTR.
   - SME duplicated triage efforts and or lacked clear direction in their investigations.
   - Low signal-to-noise ratio increased cognitive load on incident Slack channels. Lots of chatter cluttering up communications between the responders resulting in slowed or absent forward momentum.
   - Estimated indirect cost: approximately $5,000 per incident in lost productivity, responder frustration and stress.

4. After understanding the financial impacts, the team looked to Responsonomics principles to implement a strategy for specific marginal improvements. RayneCloud then adopted Responsonomic principles across three phases of the incident response lifecycle:

   a) Evaluation and Optimization of Incident supply

- Deployed feature flags to reduce change-induced volatility, reducing the risk of failure in production.
- Hardened pre-production testing to limit fragile releases that result in self-inflicted incidents.
- Developed clear SEV levels across all business units to differentiate between high severity incidents and events that do not result in significant customer or internal BI.

b) Basic Demand Modeling for high severity incidents

- Heat mapped incident frequency by geography and time of day and adjusted staffing for anticipated peaks and valleys of incident activity.
- Shifted to a tiered response model (IC + dispatched SMEs, not "all hands").
- Developed shorter on-call rotations and added additional staff to cover illness, vacations, occurrence of multiple incidents, etc.
- Empowered Incident Commanders to release anyone attending an incident not deemed to be mission critical.
- Established structured communications update cadence via a Communications Liaison.

c) Identification of labor and OC optimization opportunities

- Revised existing SEV levels to be consistent across the company. Linked SEV levels to communications protocols and executive involvement.
- Rolled out predictable, automated IC assignment and role tracking via Slack Bot.
- Created processes to understand financial and reputational BI as it pertains to allied business units such a legal, sales, etc.
- Created a financial impact model to standardize the calculation of hourly labor burn rates per SEV level.

5. RayneCloud also commissioned an external consulting firm to conduct a Gap Analysis to Best Practices study.

Using internal employee interviews to better understand the gaps, overlaps, strengths, weaknesses and opportunities, RayneCloud obtained a clear picture of their current response practices as they relate to other World Class IT incident response teams. In addition to the deep dive driven by the Senior Leadership team, the comprehensive written report and roadmap helped RayneCloud better focus their efforts on improving operation maturity. The highlights of the report included the following:

1. Time spent to understand and fine tune incident supply through improved observability has a positive impact on the demand for responders and the time they spend to resolve issues.
2. Time ≠ Cost. Reducing MTTR is a goal but doesn't always reduce financial impact.
3. Dispatching the right responders in a timely and predictable. manner and holding them accountable will reduce MTTA and have a positive impact on MTTR metrics.
4. Incident Command and coordination = cost control. Defined roles, fewer responders, and less people attending incidents due to poor communications may save ~$624K/year.
5. Quantifying responder time and effort in economic terms provides good visibility into developing business cases for staffing, tools, etc.
6. Understanding OC such as the missed onboarding for two $50K annual clients = $100,000 and the postponed data partnership integration (risking market advantage) = $50,000 can bring into sharp focus the BI of an outage.

Within six months of addressing the items in the Gap Analysis and its own findings, RayneCloud found the following improvements:

- Labor efficiency improved. SEV-1 incidents required an average of 7 responders instead of 12.
- Decision latency decreased. Clear IC authority reduced wasted minutes in escalation.
- Economic visibility increased. Incident cost sheets became part of executive reporting, reinforcing accountability at all levels.
- Trust was preserved. Customer churn triggers dropped from 12 per month to just 3.
- After linking customer billing systems to customer support systems and incident observability systems, the senior leadership team obtained a holistic economic and BI picture of an incident based on incident supply.

9 METRICS, MODELS AND MEASUREMENT

Measure what costs money; model what's important to optimize

In Responsonomics, metrics and *measurement* are not about random *data* collection and reporting for reporting's sake. One of the shortcomings of traditional uptime dashboards or service level reports is that they only show what happened in technology performance terms. Economic metrics go further: They show what it cost and what you should do next – what may or will happen. There's an adage in data collection and metrics that goes like this: "You can't manage what you don't measure." And when it comes to metrics in IT incident response, there is a huge opportunity for improvement. We seldom run into companies that really put elbow grease behind establishing useful metrics beyond assembly and resolution times, then actually take action based on the results. When data from established metrics are collected, it's fairly rudimentary information typically centered around Incident start time, SEV level, MTTA, MTTR, Incident end times, and perhaps the number of responders that attended the response. All useful for sure, but what do those metrics really mean when it comes to establishing their value from a Responsonomics point of view? As you've read in earlier chapters, additional metrics such as OC, labor calculations, etc., in addition to those metrics, help develop a much richer picture of a single incident or a batch of incidents over time. This chapter will delve into some of the rationale behind collecting more holistic metrics and how they can add value to the pursuit of improving the efficiency of incident management and incident response.

Why Metrics and Measurement Matter

Metrics in Responsonomics are more than performance indicators—

they are economic instruments. With metrics, leaders can quantify trade-offs, evaluate investments, and design strategies that maximize return on scarce resources. Every company, no matter what its size, operates in a resource constrained environment. Executives make decisions every day on allocating capital, time and resources to maximize shareholder value. Without proper metrics, financial and other types of decisions that should be data driven are made without appropriate information.

> *Key Thought:* Leaders constantly face allocation decisions—whether to invest in additional headcount, automation, or improved observability. Economic metrics bring clarity to these choices. If data shows that senior engineers spend 40% of their time chasing false positives, the return on investment for automated alert tuning becomes obvious. Instead of relying on intuition, leaders can channel resources toward interventions with the highest measurable impact.

Many of the most damaging expenses in incident response never appear on a company's income statement or balance sheet. Frustration, rework, and lost opportunity silently accumulate in the background. Metrics expose these hidden drains on productivity, reframing them as measurable liabilities instead of "just part of the job."

Metrics also function as a learning mechanism. Every incident is an opportunity to improve the system, but only if lessons are captured, measured and implemented and if they lead to systemic improvements across the entire incident lifecycle. For example, if postmortem analysis reveals that delayed escalation consistently adds 30 minutes to high-severity incidents, a metric such as MTTA can highlight the recurring pattern. Over time, tracking this measure allows teams to validate whether changes in on-call paging protocols or staffing models reduce assembly delays. Metrics convert human and technology actions into viewable trends across time, informing the cycle of continuous improvement. Finally, metrics also provide the evidence necessary to justify investments in resilience. Business leaders are rarely persuaded by purely technical arguments for redundancy, automation, or training. But when metrics translate those investments into reduced downtime, hours preserved, or churn avoided, all of which translate to dollars saved, the conversation shifts. In other words, convert operational metrics into financial metrics, so they can be analyzed side by side with all the other financial metrics of the business. Building a business case that shows that a new dashboard configuration reduced alert noise and saved $250,000 in reclaimed engineering hours is more compelling to an executive audience than a chart of CPU utilization.

Technical vs. Economic Observability

There is the critical difference between technical observability and economic observability. In short, technical observability is about IT system health. Economic observability is about business health. While a system may look stable in technical dashboards, it may also be silently draining organizational value through hidden inefficiencies. Only when metrics illuminate these costs do leaders gain the visibility to make informed, economically rational decisions.

The following hypothetical case illustrates the point:

A large online retailer discovered that its operations team was responding to more than 15,000 alerts per quarter, with fewer than 20% requiring action. At first glance, the noise seemed like an unavoidable nuisance. But when the company applied Responsonomics-style measurement, the hidden costs became visible:

- 12 senior engineers were losing an average of 6 hours per week to false positives.
- At a fully loaded labor rate of $150 per hour, this equated to $562,000 annually in wasted effort.
- More damaging, those same engineers were pulled away from planned projects, delaying a critical checkout optimization

feature expected to increase revenue by 3% in the holiday season.

The OC of delayed innovation dwarfed the direct labor expense. Armed with this data, leadership approved an investment in observability tooling and automated alert suppression. Within six months, actionable signal quality improved by 40%, and the team recovered nearly $500,000 in effective engineering capacity. The lesson was clear: Until all the obvious and more subtle costs were measured, they were invisible. Once quantified, they became undeniable—and actionable—all great content for a strong business case.

From Metrics to Economic Models

Metrics by themselves describe, but models explain and predict. The financial impact *model* seen earlier in the book illustrates this point exactly. It allows a company to take a much more objective and predictive approach to understanding the relationship between 'what we do' and 'how it impacts the company as a whole.'

> *Key Thought:* A model is any structured framework that integrates multiple metrics into an explanatory or predictive narrative about the economic impact of incidents. Models connect metrics together, revealing cause-and-effect relationships that show that incidents create financial outcomes, and how changes in behavior, tools, or staffing will alter those outcomes.

Responsonomics relies on models to connect raw measures into actionable economic narratives. This shift—from counting operational numbers to telling financial stories—is what allows leaders to design, justify, and sustain improvements. As you saw earlier in the book, a single metric, such as MTTR, is informative but incomplete. Metrics provide snapshots—models provide context.

In Responsonomics, the power of modeling lies in connecting the dots across metrics to tell a coherent economic story. For example: A company uses a notification tool to page specific technical resources about alerts in the technology stack. A review of the notification tool logs finds that the engineering team receives 10 false pages per week. These are either false positives or found to be events rather than incidents. Those 10 pages take on average two hours of engineering time which equals 80 hours per month. Assuming a junior engineer's hourly rate is $150.00, that 87 hours of time equals $12,990 per month of engineering time chasing false alarms.

If the company can drop the false page by 25%, or 2.5 pages per week x 52 weeks, it results in 130 less pages per year which in turn becomes an annual savings of approximately $39,000 of engineering time. 130 less pages per year (at 2 hours per page) allows the engineering team to recover approximately 260 hours of normal work time.

Assuming 2,080 hours of productive work hours per year, (based on 40 hours/week x 52 weeks) the engineering team in effect picks up almost 12% of a full time employee without adding headcount. This potential savings could be used for an automation justification, or some other tooling investment, or as the economic justification for better observability or revising SEV levels or some other investment in reducing the supply pipeline of incidents.

These are not abstract statistics—they are data driven, economic arguments that speak the language of business, to business decision makers. When framed this way, technical improvements become investments with measurable returns rather than operational metrics. Without metrics and models, organizations risk drowning in a sea of anecdotal information with no actionable tactics. With models, every data point contributes to a financial story that justifies investments, forecasts outcomes, and aligns operations with strategy.

Key Characteristics of a Model

By translating technical health into economic health, models align the operational elements of the business with the goals of business leadership. This alignment is essential in Responsonomics, because it ensures that incident response is not treated as a necessary evil, or a cost center, but as a strategic lever of value protection. Here are four different types of models that may assist you developing your own approach to compiling a compelling metric and model–based business case.

- Integrative: Combines raw measures into a coherent economic story.

  Example: "40 wasted hours per week × $150/hr. = $6,000 → equivalent to one senior engineer per month → justification for automation."

- Explanatory: Shows why outcomes occur, not just what happened.

  Example: Linking high false page rate → responder fatigue → slower triage → higher customer impact costs.

- Predictive: Allows leaders to forecast future scenarios.

  Example: "If false alerts decrease by 25%, responder efficiency increases by 200 hours annually, saving ~$300,000."

- Decision-Oriented: Provides evidence for resource allocation or investment.

  Example: "An observability investment of $400,000 yields a modeled savings of $1.8M in reduced false alerts and churn avoidance."

Forecasting and Scenario Analysis

Economic models extend beyond today's cost to forecast tomorrow's outcomes. For example, leaders can build models that show:

- The financial effect of reducing deployment-related incidents.
- The projected savings from lowering average time-to-detection.
- The long-term ROI with improved observability tooling.

This approach mirrors financial scenario planning used in business strategy and has a better chance of appealing to key business and financial stakeholders. It allows the incident response team leader to create savvy and compelling narratives to incrementally improve the odds of operational success. Just as a sales team models revenue forecast under different market conditions, incident response leaders can model operational costs under different reliability strategies. As you've seen so far in this book, incident response should be viewed as a microeconomy inside the macroeconomy of the entire company.

Here's an example of how building a model can help clarify the collected metrics:

A global SaaS provider tracked the false dispatch rate across 20 service teams. Initially, the metric was reported as "65% of alerts required no action." While useful, the number lacked urgency. When the operations team built a model calculation, the picture changed:

- 65% false page rate → 18,000 wasted hours annually
- At $140/hour average loaded cost → $2.52 million in avoidable expense
- Modeled an estimated reduction of false alerts by 30% → savings equal to 9 full-time engineers

Armed with this model, leadership approved a six-month automation initiative. The project cost $400,000 but delivered modeled savings of

$1.8 million within the first year. The narrative shifted observability from a technical upgrade to a cost-saving investment.

In most cases, a company budget does not have a line item called “Inefficiency,” and without models, some real costs may remain invisible. Responsonomics shines a bright spotlight on inefficiency so that once it’s recognized, action can be taken. Models also give leaders permission to act and take financial and operational risks based on data rather than hunches or things that sound like good ideas.

Why Models Matter to Decision-Makers

Executives and Boards may not act on simple operational metrics alone. What drives decisions is a compelling economic narrative that accompanies those metrics. A chart showing CPU spikes may not inspire a CFO to approve a budget request, but a model showing how those spikes translate into $2 million in lost productivity and churn risk certainly will. The technical measure has been reframed as a business consequence—and now it competes for attention and resource allocation alongside marketing campaigns, sales targets, and capital investment projects. Again, the end game is to create alignment between the incident response system and business leadership, in terms the business understands.

Consider another example: A financial services company tracked recurring outages in its transaction processing platform. Initially, the operations team reported this as “MTTR has improved by 15% over the last quarter.” While technically accurate, the message did not resonate at the executive level.

When reframed as an economic narrative, the picture changed:

- Technical metric: 15% MTTR reduction
- Economic impact: Faster recovery equated to 2,400 fewer failed transactions per month, preserving $1.2 million in retained revenue and reducing customer service costs by $300,000

With this framing, leadership better understood the impact of a 15% MTTR reduction and began routinely tracking economic and BI alongside technical metrics in quarterly reviews.

The Measurement Pyramid: Building Data Into a Strong Business Case

Responsonomics frames the action of metric measurement as a multi-level pyramid, with each layer building on the one below it. This structure emphasizes progression as leadership moves from collecting raw observations to implementing economic based decision-making. The pyramid ensures leaders climb from observation → insight → action rather than looking at metrics in isolation or drowning in dashboards full of orphaned information and vanity metrics.

Level 1: Data

At the foundation of the pyramid lies data—the raw material of measurement. This includes system logs, timestamps, traces, event records, monitoring outputs, labor calculations, OC approximations, etc. By itself, raw data has limited meaning. Thousands of log entries or alert timestamps may capture system behavior, but without interpretation, they remain noise rather than knowledge. Many companies struggle with establishing metrics and data collection, which is ironic when such a large part of technology relies on data to drive the bottom line of the business. Data is abundant, cheap, and often overwhelming. Its value emerges only when transformed into metrics.

Level 2: Metrics

The next level is metrics, which transform raw data into standardized, comparable measures. Metrics are whatever facets of the incident lifecycle you want to focus on and can include values such as:

- Number of incidents
- False page rate

- Change failure rate
- MTTA
- MTTR
- Labor cost per incident
- Linkage between incidents and financial impacts to sales
- OC
- BI

Metrics allow organizations to identify patterns, benchmark performance, and highlight areas of concern. However, metrics alone remain simply descriptive. Again, they show what is happening, but not why it is happening or what it means economically. The important part in establishing metrics is consistency. While the data that feeds the metric may not be perfect, it will show you trends over time.

Level 3: Models

The third level is where Responsonomics takes shape. Metrics—and the data that feeds them are useful—but they only describe past events. Models build narratives that explain and predict, helping leaders see down the road and make value judgements on systemic improvements or the state of the status quo.

This step up the pyramid is where you apply the predictive lens that allows leaders to compare investments in resilience with competing initiatives across the business. It's the place where context is built around the mixture of metrics. There is no one standard model that applies to all situations; it is up to you to determine how best to frame your business case. Consider the following examples:

Example 1: Slow Escalation Impact Model

Purpose: Estimate the economic cost of delayed escalations during high severity incidents.

Inputs (data and metrics):

- High-severity (SEV-1/SEV-2) incidents per month: 15
- Average delay due to slow escalation from Level 1 to Major Incident Management Team: 45 minutes
- Average number of engineers affected per incident during delay: 3
- Average loaded labor rate per engineer: $160/hour

Model Logic:

- 45 minutes = 0.75 hours of delay per incident
- Labor impact per incident = 0.75 hours × 3 engineers = 2.25 hours
- Total monthly impact = 2.25 hours × 15 incidents = 33.75 hours/month
- Annual impact = 33.75 hours × 12 months = 405 hours/year
- Total labor cost = 405 hours × $160/hr. = $64,800/year

Predictive Scenario:

- If escalation speed improves and reduces delay by 50%,
  → time saved = 202.5 hours/year
  → cost savings = 202.5 hrs. × $160 = $32,400/year
  → Equivalent to reclaiming ~10% of a full-time engineer (FTE)

Responsonomic Value: This model demonstrates how reducing handoff latency and improving escalation protocols yields measurable returns—not only in faster time-to-resolution but also in engineer availability, reduced burnout, and higher project velocity. Framed this way, investing in escalation training, real-time routing tools, or tiered engagement automation becomes a justified business expenditure rather than an optional process improvement.

Example 2: MTTR Improvement and Revenue Preservation Model

Purpose: Forecast financial benefit of reducing mean time to resolution (MTTR).

Inputs (data and metrics):

- Current MTTR = 90 minutes
- Target MTTR = 70 minutes (20 min improvement)
- Transactions lost per minute of downtime = 200
- Average revenue per transaction = $2

Model Logic:

- Current downtime loss per incident = 90 × 200 × $2 = $36,000
- Improved downtime loss per incident = 70 × 200 × $2 = $28,000
- Savings per incident = $8,000

Predictive Scenario:

- If 30 major incidents/year → $8,000 × 30 = $240,000 annual revenue preserved

Responsonomic Value: The model reframes MTTR not as minutes saved but as revenue preserved—a language executives understand.

Example 3: Automation ROI Model

Purpose: Predict ROI by automating triage activities.

Inputs (data and metrics):

- Average triage time per incident = 15 minutes
- Number of incidents/month = 200
- Responder loaded labor rate = $125/hour
- Automation cost = $400,000 (one-time)

Model Logic:

- Triage labor = .25 hr. × 200 incidents × $125 = $6,250/month
- Annual triage labor = $6,250 × 12 = $75,000/year

Predictive Scenario:

- If automation reduces triage labor by 80% (75,000 x 80%) annual savings = $60,000
- Payback period = $400,000 ÷ $60,000 ≈ 6.7 years
- If incident volume grows 20%/year, savings compound to reduce payback to ~5 years

Responsonomic Value: The model demonstrates that automation is not just a technical upgrade, but an ROI-based investment decision.

Level 4: Decisions

Decision making tops the measurement pyramid; it is the point where measurement becomes actionable. Data and metrics reveal patterns; models provide explanation and foresight. Decisions transform measurement into economic value. A decision is the tangible evidence that measurement occurred, and a change initiative has begun. In business, decisions should not be made on intuition alone. They are guided by evidence-based economic arguments derived from models. Leaders use these arguments to weigh trade-offs, allocate scarce resources, and prioritize interventions that yield the highest return.

The Graphic in Figure 9-1 depicts the four levels in the measurement pyramid.

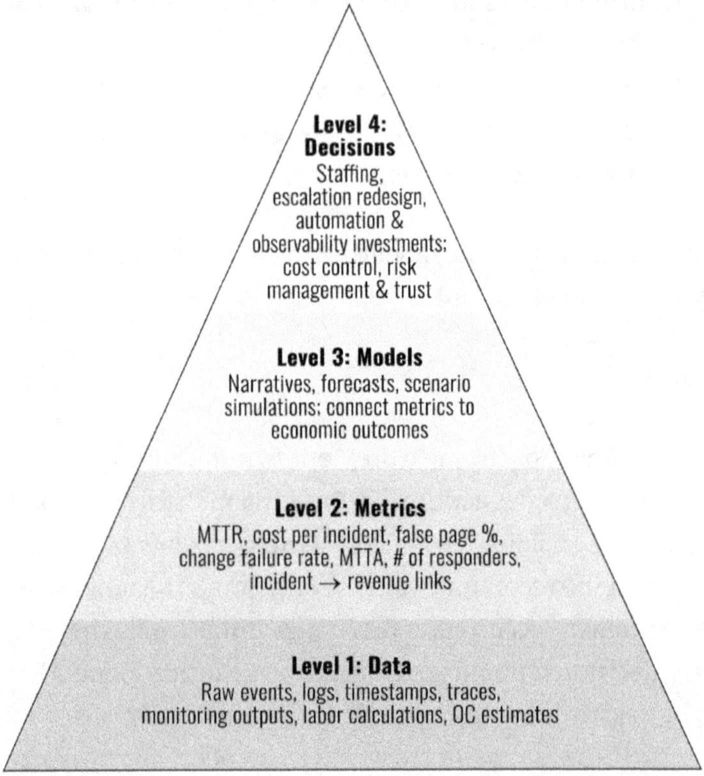

The Measurement Pyramid (Responsonomics): progression from observation → insight→action. Avoid dashboard theater by elevating data into decisions.

Figure 9.1

Perhaps the most visible application of Responsonomic decision-making is in technology investment. Automation and observability often require significant upfront costs. Without models, they may be perceived as nice-to-have improvements. With models, they become ROI-backed business cases. The following examples illustrate how data, metrics, and models translate into actionable, ROI-backed conclusions. Each statement represents a decision that shifts incident management from a reactive cost center to a strategic engine of value protection. Anchoring each conclusion in cost, time, or trust increases the likelihood that it resonates with executive decision-makers. The following examples of decision making action statements are general but reflect an executive level summary of the project, initiative, or action. These statements serve as the opening statement to the business case being made. Certainly, there would be accompanying information to each initiative.

Technology Investment Decisions

- Automation in triage will reduce MTTD by 10 minutes, avoiding $1.5M annually in customer churn.
- Observability upgrades will cut false alerts by 40%, equivalent to gaining 6 FTEs.

Staffing and Escalation Decisions

- Reducing escalation to senior engineers by 25% will save $820,000 annually in high-cost labor, and will free 4 senior staff for strategic initiatives.
- Shifting on-call coverage to a global follow-the-sun model reduces MTTA assemble by 15 minutes, preserving $600,000 per year in avoided downtime.

Efficiency and ROI Decisions

- $300,000 invested in alert suppression tools pays back in under 9 months by eliminating $1.2M in wasted Labor from false positives.
- Automated postmortem reporting reduces analyst effort by 80 hours per month, equivalent to $144,000 annually, and accelerates learning cycles across teams.

Customer Impact and Trust Decisions

- Improving detection accuracy by 20% prevents incident recurrence, reducing repeat outages by 50 cases per year—valued at $2.4M in avoided customer support costs and lost revenue.

- Automation in triage will reduce MTTD by 10 minutes, avoiding $1.5M annually in customer churn.

When expressed this way, investment decisions shift from technical debates to financial justifications that resonate at the executive and Board levels. At this level, measurement is no longer about reporting uptime or tracking incident counts. It becomes a strategic engine for cost control, risk management, and trust preservation.

Responsonomics positions decision-making not as the end of measurement, but as its purpose. Metrics and models have no intrinsic value unless they guide action. When organizations consistently use economic models to inform staffing, escalation, and investment, they transform incident response from a reactive burden into a proactive system of value protection.

Leading vs. Lagging Indicators

Measurement is most valuable when it not only explains the past but also shapes the future. In Responsonomics, this means distinguishing between *lagging indicators* and *leading indicators*.

Lagging indicators are retrospective measures—they describe outcomes after they occur. Common examples in incident management include:

- MTTD
- MTTA
- MTTR
- Incident Count: The number of events logged in a given period.
- Customer Downtime Hours: The total outage time experienced by end-users.
- Escalation Frequency: How often incidents require hand-off or elevation, potentially reflecting skill gaps or misaligned roles.
- Reopened Incidents: Number of incidents that recur or are reclassified due to incomplete resolution—indicates potential technical debt.

- SLA Breaches: Incidents that exceed response or resolution service level agreements, often triggering contractual or reputational consequences.
- Support Cost per Incident: Total financial burden incurred per incident, aggregating labor, tooling, overtime, and OC.

While useful for trend analysis and historical reporting, lagging indicators provide little foresight. They tell you how well (or poorly) you performed certain tasks like assembling the team or the length of time taken to resolve some number of incidents. While they are descriptors of past performance, they certainly feed into the financial impact model. As an example, if a company chooses to pursue better MTTA metrics, and tracks those numbers over time, there may be tangible improvements in MTTR metrics. If we assemble quicker, we should, in general, resolve quicker. Or, if we improve observability, it should reduce the numbers of incidents that the team responds to, which in term may lead to a lower overall price tag for overall incident response. Lagging indicators are a great way to implement and measure this form of logic: If we make changes and improvements to X how will that impact Y? This rings true all through the incident lifecycle.

> *Key Thought:* Lagging indicators measure what happened yesterday. Leading indicators reveal what will happen tomorrow. Responsonomics insists that organizations weigh both but place priority on signals that offer foresight, enabling proactive investment and strategic resilience.

The financial impact model then is the tool to express the economic impact of the improvement.

Here's a hypothetical scenario illustrating why lagging indicators (when used singularly) have shortcomings:

At a global SaaS company, leadership tracked lagging indicators religiously. Dashboards highlighted MTTR improvements and a steady incident count, suggesting stability. Within six months, turnover spiked among senior incident responders, and customer complaints increased sharply.

A postmortem revealed the warning signs had been present all along:

- Paging volume had risen by 35% over two quarters
- False page rate exceeded 60%, eroding focus and morale
- Engineers reported burnout in internal surveys, but this data was never elevated as a metric

Because leadership prioritized lagging measures, they missed the leading signals of cost escalation. The result was higher attrition costs, delayed product releases, and a measurable decline in customer trust. When the company shifted to a Responsonomics approach, they began tracking leading indicators—alert noise, escalation debt, and staffing load. Within a year, false alerts dropped by 40%, attrition stabilized, and economic models demonstrated $1.5 million in regained engineering capacity.

In Responsonomics, leading indicators are treated as predictive metrics—early-warning signals that reveal rising costs, declining trust, or approaching organizational strain before these outcomes fully materialize. By tracking these signals, organizations can make proactive, economically sound decisions. Leading indicators can reflect the human and organizational dimensions of incident response. Examples of leading indicators that reflect human dimensions include:

- On-call burnout patterns: Upward trends in alerts, especially during weekends or overnight hours, signal an unsustainable workload and predict future absenteeism or turnover.
- Sick leave spikes: A sudden increase in sick days among engineers—especially following major incidents—is a red flag for psychological fatigue and decreased long-term availability.
- Exit interview mentions: Repeated comments about "alert fatigue," "escalation stress," or "poor handoffs" in exit

interviews point to systemic friction that may not appear in formal metrics.

- Decline in volunteerism: When senior engineers stop volunteering for high-severity responses or mentoring new responders, it signals quiet disengagement and rising OC.

Technology leading indicators reveal patterns in systems, tooling, or workflow that forecast increased risk or inefficiency:

- Surging non-actionable alerts: A growing percentage of alerts that require no action indicates a poorly calibrated monitoring system and wasted engineering effort.
- Delayed escalation timestamps: If Tier 1 or bot-based escalations show increasing lag before human pickup, it suggests a growing gap in organizational response readiness.
- Repeat incident patterns: Clusters of incidents tied to the same service or failure type show that root causes are being patched—not resolved—draining time and trust.
- MTTD drift: A rising Mean Time to Detect, even by a few minutes per month, can quietly erode performance SLAs and customer experience, long before an outage occurs.

Within a Responsonomic framework, leading indicators function not merely as operational signals, but as forward-looking economic forecasts. They reveal how today's conditions will shape tomorrow's costs. Most critically, these internal inefficiencies eventually surface externally, degrading customer trust as response quality, consistency, and communication suffer.

Prioritizing leading indicators allows organizations to intervene before these downstream costs fully materialize. By monitoring early signals—such as alert quality, on-call load distribution, and escalation frequency—leaders can make preemptive investments that alter the economic trajectory of incident response. This may include strengthening staffing models, tuning alert thresholds, redesigning escalation paths, or expanding targeted automation to absorb predictable workload. Acting at this stage is far less costly than addressing the consequences later, when burnout, attrition, and reputational damage have already taken hold.

Responsonomics emphasizes leading indicators precisely because they create this bridge between present signals and future costs. They transform incident management from a reactive discipline focused on yesterday's outcomes into a proactive economic practice, enabling leaders to manage not just incidents themselves, but the financial and human systems that surround them.

Consider an on call team that receives an average of 40 alerts per week, more than half of which are non-actionable false pages. Individually, each alert appears trivial—a few minutes to acknowledge, a quick check, and a dismissal. Economically, however, the pattern tells a different story. Engineers begin to expect interruptions during evenings and weekends, fragmenting recovery time and reducing cognitive resilience. Over several quarters, response times to real incidents lengthen, not because systems have degraded, but because responders are fatigued and disengaged. Informal workarounds emerge as engineer's mute alerts or delay responses, increasing risk exposure. Meanwhile, improvement work stalls as capacity is consumed by noise, and leadership misinterprets stable incident counts as evidence that the system is "under control."

By the time a major outage occurs, the true cost of the false-page rate becomes visible. Resolution is slower, communication is uneven, and customer confidence erodes. What appeared months earlier as a minor operational irritation is revealed, in retrospect, as a leading indicator of labor inefficiency, burnout, innovation slowdown, and trust degradation. A Responsonomic approach may have justified earlier investment in alert tuning, workload redistribution, or automation at a fraction of the eventual cost.

Avoiding Metric Theater

Measurement becomes dangerous when it devolves into metric theater—the practice of tracking or showcasing numbers that appear impressive but fail to truly inform decisions or form the basis for specific marginal improvements. Vanity metrics create the illusion of progress without improving economic outcomes. This risk is significant: When organizations celebrate the wrong signals, they unintentionally reward behaviors that increase hidden costs and put operational maturity at risk.

Metric theater often takes the form of dashboards that highlight easily improved, low-value measures. Examples include:

- Low page volume celebrated, even though end-user downtime is increasing
- MTTR improvement touted, while OC remain unmeasured
- Faster recovery times rewarded, even when fixes are unsustainable and recurrence risk grows

In each case, the numbers look favorable, but they do not reflect the true economic reality of incident response.

In many cases, vanity metrics can result in the following:

- False Confidence: Teams believe they are improving when, in fact, costs are shifting elsewhere (e.g., quicker fixes but more frequent rework).
- Misallocation of Resources: Leaders invest in areas that move the "dashboard needle" but fail to reduce long-term costs.
- Erosion of Trust: f stakeholders or employees discover that celebrated improvements are hollow, confidence in engineering and leadership declines.

Consider this scenario to summarize the pitfalls of vanity metrics:

A mid-sized cloud provider proudly reported a 25% improvement in MTTR across its quarterly reviews. On the surface, this suggested a leaner, faster response capability. However, customer complaints were rising, and support costs had doubled.

A closer look revealed why:

- Engineers were applying temporary fixes to resolve alerts quickly, satisfying the MTTR metric.
- These reported quick wins led to recurring incidents, which multiplied long-term labor costs.
- Meanwhile, OC mounted as engineering hours were diverted away from feature development.

The verdict? The vanity metric of faster resolution hid the true cost trajectory. Leadership reframed measurement through Responsonomics by identifying recurring incident frequency as the real driver of costs. Investments shifted toward permanent fixes, automation, and observability. Within two quarters, incident volume dropped by 30%, customer churn stabilized, and the company's economic models showed genuine savings.

10 AUTOMATION AND LABOR SUBSTITUTION

Automation should absorb the noise so humans can manage the signal

In traditional economics, labor substitution refers to replacing human work with machines, algorithms, or processes to reduce cost and increase productivity. In Responsonomics, this principle applies directly to incident response: Every minute of human effort that can be meaningfully automated is an opportunity to reduce cost, human error, and fatigue. But automation isn't just about saving time or reducing headcount—it's about achieving optimum efficiency first and enabling scale and repeatability second. Keep in mind that any form of automation isn't inherently good or bad, effective or ineffective, useful or useless. It's a tool like anything else and must be implemented in a holistic manner for the micro and macro benefits of the company. This chapter explores how organizations can intentionally model, prioritize, and implement labor substitution through automation across the incident lifecycle, to the benefit of the company and the workforce.

The Promise of Automation in Incident Response

There is currently much discussion about automation and AI in incident response, and one thing is clear—there is no consistent viewpoint on the topic. Our goal with this chapter is to stimulate your own thinking on the matter. It is not a de facto endorsement or critique of AI in any way, or recommendation of any company, technology, or tooling. There are many compelling arguments to be made in favor and against, but the fact is that automation in the right setting for the right reasons may offer several well-documented advantages.

- Automated processes are beneficial when routine tasks (such as log analysis, alert correlation, or remediation playbook execution) are completed much faster and with fewer errors than humans.
- When replacing repetitive tasks with automation allows scarce senior engineering capacity to be redeployed toward higher-value activities like architecture, prevention, or customer engagement.
- As systems grow in complexity, automation enables incident response teams to handle more alerts, more services, and more data without proportional headcount increases.
- Appropriate automated workflows reduce the risk of human fatigue and error during prolonged or high-severity events.

On the flip side, despite its initial appeal, automation and AI are not without drawbacks:

- Over-automating without careful monitoring can lead to "silent failures," where automated responses mask problems until they escalate.
- AI-driven incident triage depends on data quality. Poor or biased data sets can reinforce ineffective or even harmful response patterns.
- Automation requires continuous updating as systems evolve. Neglected automation can become automation debt, adding complexity rather than reducing it.
- Experienced responders often detect subtle anomalies or context clues that current AI models cannot capture. Eliminating too much human oversight risks missing the unknown unknowns.
- Automated systems can experience performance degradations, service outages and security events, just like any other IT system.

AI in Incident Response: Emerging Perspectives

AI adds predictive and adaptive capabilities on top of traditional automation, extending the ability of responders to detect, analyze, and resolve incidents. Unlike scripted automation, which executes predefined tasks, AI systems can continuously learn from patterns in telemetry, human actions, and incident outcomes. When applied thoughtfully, this introduces new layers of efficiency and foresight into incident response.

Modern Incident Management increasingly leans on AI and machine learning (ML) not just for automation, but for early signal detection and anomaly prediction. In Responsonomics, this capability is central to reducing response costs, protecting labor capacity, and shortening the incident lifecycle. Traditional monitoring tools rely on static thresholds—fixed values that trigger alerts when exceeded (e.g., CPU > 90%, response time > 2 seconds). While effective in some cases, threshold-based systems are inherently reactive, brittle, and prone to false positives or missed signals. Machine learning models, by contrast, analyze patterns across multiple data streams simultaneously: logs, metrics, traces, traffic, and user behaviors. These models can establish baselines dynamically and flag subtle deviations—often identifying signs of trouble before performance visibly degrades or users report issues. Here is a simple example of how AI based automation function sin a positive manner: A financial services provider applied AI-based anomaly detection to its fraud monitoring system and discovered performance degradation two hours before traditional alerts would have fired, preventing a customer-impacting outage.

Natural Language Processing (NLP) for AAR and Pattern Recognition

NLP, a branch of AI, offers powerful tools for making unstructured incident data—such as Slack threads, helpdesk tickets, AAR, and engineering notes—economically actionable. In Responsonomics, this matters because unstructured labor activity (e.g., triage discussions, ticket reviews, and root cause analysis) often represents hidden, unquantified cost. NLP helps surface repeatable insights, reduce duplicated effort, and accelerate decision-making.

Traditionally, AAR and ticket triage depend on manual reading, tagging, and clustering. This process is time-consuming, inconsistent, and labor-intensive—especially at scale. NLP techniques can ingest and process thousands of natural language records, performing the following functions:

- Clustering similar incidents or symptom patterns.
- Extracting root causes from repeated narrative language.
- Identifying recurring keywords, entities, or affected services (e.g., repeated API failures, timeouts, user complaints).
- Mapping semantic similarity across alerts, on-call handoffs, and status updates.

When applied effectively, NLP techniques reduce the time responders and SREs spend on postmortem analysis or ticket classification. Consider this example: A company that performs 25 AAR per quarter, each requiring 4 hours of manual synthesis, spends 400 hours/year on post-incident documentation. If NLP reduces time per postmortem by 50%, it reclaims 200 hours/year—equal to $30,000+ in engineering time (assuming $150/hour blended rate).

Beyond labor savings, NLP also enhances learning velocity: Patterns are detected earlier, contributing to faster systemic fixes. This reduces recurrence rates of known failure types, shrinking both MTTD and MTTR.

One of the most powerful economic outcomes of NLP is its ability to reframe response strategy from tactical to strategic. For example:

- Instead of resolving 40 similar tickets individually, NLP clusters them as manifestations of a single recurring bug.

- Instead of treating 10 alerts as separate events, NLP maps them back to a single overloaded API or authentication failure.
- NLP enables response consolidation, faster root cause isolation, and ultimately, fewer incidents in the future.
- AI systems can augment human responders by offering likely root causes, probable blast radius, or recommended next steps.
- Instead of replacing responders, these tools shorten triage time by narrowing the field of possible causes, letting responders focus on validation and execution.

Despite these types of successes, experts caution against the AI panacea viewpoint. As Google's SRE community has noted, AI should be deployed in narrow, well-defined tasks rather than broad replacements for human judgment.

> *Key Thought:* Automation reduces workload. AI augments judgment.

As support for this thought, Gartner's 2023 incident automation report similarly emphasizes the need for a "human-in-the-loop" model to balance efficiency with safety. As the world put millions of new AI GPUs into production, AI will continue to grow and evolve, so to say we are in the early innings of this game would be an understatement.

Responsonomic Lens on Automation

Building on the principle that incidents are economic events, Responsonomics reframes automation not as a technical ambition, but as an investment decision. The objective is not to automate for its own sake—whether through AI or conventional tooling—but to deploy automation where its economic return is clearest. High-frequency, low-variance tasks are ideal candidates for mechanization, while low-frequency, high-impact situations—such as customer communications and executive escalations—should remain the domain of human judgment. Accordingly, automation initiatives should be evaluated not by their technical sophistication, but by their measurable effects on labor efficiency, cost avoidance, operational stability, and the preservation of customer trust.

> *Key Thought:* The goal of Responsonomic labor substitution is not to eliminate human responders. It is to optimize their deployment, preserving their attention for the moments that matter most: complex, ambiguous, and high-trust incidents.

From a Responsonomics point of view, human labor must be treated as both a cost to be managed and a form of capital to be preserved. As a cost, labor can be quantified in terms of direct hourly rates and OC of delayed innovation. As capital, skilled responders represent irreplaceable institutional knowledge and decision-making capacity. Burnout, attrition, or misallocation of senior talent are longer term economic costs far greater than the immediate labor cost of a single incident. Thus, automation's value is not merely in replacing tasks—it is in protecting human capital by ensuring that responders apply their expertise where it yields the greatest organizational return.

Automation in Responsonomics is not merely a tool for cost-cutting. It is an economic multiplier: A one-time investment that compounds over time, reducing toil, protecting scarce human capital, and enabling scale without proportional increases in labor cost.

Common Pitfalls and Hidden Costs of Automation

While automation promises efficiency, scale, and predictability, it is important to recognize that automation is not free... or easy. Poorly scoped or rushed initiatives can result in costs that outweigh the intended benefits. Responsonomics teaches that automation must be evaluated through an economic lens, accounting not only for savings but also for the upfront cost of engineering time to integrate and understand the technology, searching out any hidden costs or ongoing liabilities. Like many other types of tools and improvement initiatives, automation has long term maintenance, technology upkeep, and the potential long-term linkage to a critical vendor or technology. In many cases, if you hitch your wagon to particular tool or vendor, it's tough to walk back the decision once significant cash and time investments have been made.

Automation, especially in legacy systems, often introduces long-term upkeep. Scripts may break when infrastructure changes, monitoring logic evolves, or third-party APIs are deprecated. What begins as a one-time automation project can quickly become an automation debt trap, consuming more engineering hours than the manual process it replaced.

Another common pitfall is automation triggered by false positives. If observability data is noisy or alert thresholds are poorly tuned, automation may act on events that do not represent true failures. In practice, this can make outages worse—for example, automatically restarting a healthy service or applying the wrong remediation step. Such errors not only increase downtime but also erode trust in automation across the team.

Automation rollouts can stall due to complex change management processes. Regulatory and compliance requirements, coupled with conservative risk cultures, often slow down deployment. Delays can erode the ROI of automation. In some cases, engineers see the solution but are slowed by organizational friction. In other cases, executives may see system improvements with automation but are wary of implementation due to open ended up front costs and ambiguous ROI timelines. Initial cost reductions go unrealized, thereby reducing potential longer term cost savings.

Modeling the *Total Cost of Automation (TCA)*

Like any other business initiative, leaders are encouraged to model the TCA before approving any project. This economic discipline ensures automation remains a multiplier of value, not a hidden liability. TCA goes beyond the initial effort to buy or build, capturing the full lifecycle costs of automation. By accounting for development, integration, and maintenance, organizations can compare the projected cost of automation against expected labor savings or OC avoidance.

Development Time

Every automation initiative begins with an engineering investment, and this can represent a significant upfront cost. While the long-term benefits of automation and AI often outweigh these expenses, leaders must recognize that development is rarely a quick win. Properly scoping, designing, and delivering automation requires dedicated labor, infrastructure, and planning.

Typical development costs may include:

- Building Scripts, Workflows, or Runbooks.
- Writing and testing code for repetitive tasks such as service restarts, database failovers, or log parsing.
- Developing playbooks that can be reliably executed under pressure without human intervention.
- Scheduling cross-team workshops to define automation triggers (what should start it), fail-safes (when to stop it), and escalation pathways (who is alerted if it fails).
- Budgeting time from senior engineers and architects to align automation with existing incident management policies.
- Preparing and testing clear runbooks, wikis, or training modules to ensure automation is understood and trusted across teams.
- Investing time in peer reviews, demonstrations, and onboarding sessions.
- Connecting automation to monitoring tools, ticketing systems, and messaging platforms.

- Anticipating API work if vendor systems do not integrate cleanly.
- Building test environments that simulate realistic incident conditions.
- Running load, stress, and fault-injection tests to validate that automation performs correctly at scale.
- Ensuring automated actions comply with regulatory requirements (e.g., in healthcare, finance, or government).
- Conducting penetration testing and security audits to prevent automation from becoming a new attack vector.
- Hiring or contracting engineers with expertise in automation frameworks or AI/ML model development.
- Providing training for existing staff to close skill gaps.
- Procuring datasets for training and validating AI-driven automation.
- Accessing cloud GPU/TPU resources for model training.
- Budgeting for ongoing expenses for model tuning and retraining as systems evolve.

While development time is the most visible cost of automation, it is often underestimated. An organization that expects to save labor time through automation must first account for engineering hours required to build, integrate, test, and validate the solution. From a Responsonomics perspective, these costs form part of the TCA and must be weighed carefully against the projected labor and opportunity savings.

Also, automation rarely if ever operates in isolation. Integration across systems, services, and monitoring platforms often takes as long—or longer—than initial development. For example, even just two weeks of quality assurance (QA) and integration testing at $125/hour may add $10,000 to the project's TCA. In complex organizations, QA integration and testing may take significantly longer, may involve a greater number of engineers, or may require support from a broader number of vendors. Automation is definitely not a "set it and forget it" solution. As systems evolve, runbooks and scripts require continuous upkeep.

Consider the scenario below:

A fast-growing company faced recurring manual processes during high-traffic periods, including scaling servers, performing database failovers, and clearing cache clusters. Each task was low risk but repetitive, consuming more than 200 engineering hours per quarter.

Leadership faced a choice:

- Hire two additional engineers to keep pace with demand (annual cost ≈ $300,000).
- Invest in automation runbooks to replace the top 10 recurring tasks (one-time development cost ≈ $50,000).

The decision was reframed through a Responsonomics model:

- Automation investment ≈ $50,000
- Labor savings in year one ≈ $240,000
- Payback period ≈ 2.5 months

The result was transformative. The team avoided headcount growth, reduced on-call fatigue, and freed senior engineers to focus on resilience architecture. Most importantly, automation scaled operations to handle 2× system growth without 2× labor cost, a direct example of how automation provides a force multiplier to the capacity of the most expensive people.

> *Key Thought:* Not all tasks are worth automating. By modeling TCA, organizations avoid automation debt and ensure resources are directed where they yield the highest economic return. When applied rigorously, TCA analysis transforms automation from a technical preference into a strategic investment decision.

Automation only delivers positive ROI when total lifecycle costs are measured and justified. Poorly chosen automation projects can create more cost, fragility, or reputational risk than they save. To summarize this point, consider another TCA evaluation example:

A large e-commerce company invested heavily in automating order-processing. The automation took three months to develop, costing approximately $200,000 in engineering time. However, because the underlying legacy system frequently changed, the automation broke repeatedly requiring ongoing patches and developer support.

Within 18 months, the company had spent an additional $300,000 in maintenance, while labor savings were only $250,000. In hindsight, leadership recognized that this process was a poor candidate for automation, given the instability of the legacy environment.

To help navigate the minefield of automation implementation considerations, we recommend that technical operational leaders think like executives before they present the business case for new investment to executives. In other words, use the language of business—including accumulated TCA and anticipated labor savings—to make your case for IT automation. Ask yourself this question: If someone else was making this pitch to you, would you find this investment attractive? We offer these diagnostic questions to help stimulate your own thinking on automation opportunities:

- How often does this process occur?
- Will automation savings compound over time, or is usage too rare to justify the investment?
- Is the process logically straightforward, or does it involve ambiguous decision paths?
- Can automation handle edge cases reliably, or will humans still be required for exceptions?
- What is the potential harm if the automation fails or misfires?
- Would errors damage customer trust, regulatory compliance, or

revenue streams?

- What are Plan B and plan C should the automation fail?
- Does this process require strategic judgment, negotiation, or empathy that automation cannot replicate?
- Does automation risk diminishing institutional learning by removing humans from critical decision cycles?
- What is the estimated TCA (development, integration, maintenance)?
- Do projected savings exceed TCA within a reasonable payback period (≤ 12–18 months)?
- Does automating this process advance broader business and resilience goals?
- Will this automation free talent for higher-value strategic work, or simply shift effort elsewhere?

In summary, it is safe to say that automation is a tool for consistency and speed, not a replacement for care, strategy, or judgment. There are times when a process just doesn't lend itself to automation or can't be justified to reduce headcount. And as any experienced manager knows, once headcount is gone it's quite difficult to get it back. To that end, assess overall efficiency first, and if headcounts can change after that, then it may be a wise choice. You can avoid being seduced by the notion of automation by applying a structured economic and strategic approach whereby automation becomes an asset, not a liability.

11 OPERATIONAL MATURITY

Operational Maturity begins when organizations start owning their response system's economics

Operational Maturity—the subjective and objective assessment of how efficiently Incident Management and incident response is carried out in an organization— is developed over time, often unevenly, and usually with challenges along the way. Just as individuals learn through experience, organizations evolve through cycles of trial and error, failure and improvement. The Operational Maturity Model (OM), developed by Blackrock Partners, provides a structured way to understand this progression. And while it appears toward the end of this book, in reality and practice, it is the starting point for any leader interested in a Responsonomics approach to process improvement. To begin the journey of getting better, it is vital to understand the state of the response system in its current form. The OM, like any other model, is a way to express a situation in concrete, easily understood terms.

The Blackrock OM model identifies five phases that most organizations travel through, each with distinct strengths, weaknesses, and economic signatures. Advancing from Phase 1 to Phase 5 is not simply about compliance or process—it is about how effectively an organization dives into the details of the incident lifecycle as it applies to the company's business goals and objectives. Each phase represents a higher degree of control, predictability, and economic efficiency. The OM model is included at this point in the book because by now, you should have a firm grasp on how your current incident response program functions and some ideas of where the shortcomings may lie. Putting a numeric value to the status quo sets you up to track the progression of improvement(s) in a more quantitative way. Each phase has unique characteristics, and your incident response program may not fit neatly into any particular phase. However, it does provide a starting point, and targets for advancing the company's operational maturity. An example of the Blackrock OM model is shown in figure 11.1 (yes, we recognize the image may be hard to read! We will show more detail in the remainder of the chapter). To access the full survey, and receive free customized results specific to your company, visit our website at Blackrock3.com.

	Phase 1	Phase 2	Phase 3	Phase 4	Phase 5
Maturity State	**Initial**	**Managed**	**Defined**	**Data Driven**	**Optimized**
Descriptive State	**Reactive**	**Responsive**	**Predictable**	**Repeatable/Scalable**	**Sustainable**
Incident roles	• No formalized incident response (IR) training • No clear roles and responsibilities for incident management • Incident communications to key stakeholders and customers is not formalized with a recognized role	• Some level of internal on boarding for incident response • Key incident response roles implemented to some level • No clear and efficient linkage between resolution and key stakeholder communication	• Command staff functions recognized, implemented and supported by leadership • All key responders and SME's trained to a consistent standard • Regular and predictable briefing/comms cadence	• Ongoing training and exercises for all key responders • Large scale incident management in place where applicable <<Unified Command>> • Dedicated communications function assigned	• Clear plans to recruit and replace team members. • IR team may reach out to customers/key clients/other business units to assist with building joint response capabilities
Processes	No documented process for dispatch, resolver engagement, resolution or After Action Reviews	• Some level of formalized dispatch process and SLA's for key responders. • Monitoring tools integrated into situational awareness. • Blameless After Action Reviews may be completed, but not integrated in Q/A & Q/I	• On call rotations predictable and key responders assemble quickly. • Playbooks and Standard procedures developed • Accountability for performance and responder duties is clear to all	• Full support of the end to end IR process from senior leadership. • All responders accountable for performance	IR is accepted as an integral part of defending the business against financial loss, reputational risk, and loss of customer trust
Engagement	Company does not recognize or support incident response as an entity or discipline. Best efforts are relied upon from individual contributors	Mean Time to Assemble (MTTA) is unpredictable. Dispatching tools, process and accountability in place, but inefficient or outdate or inconsistently used	MTTA is optimized and repeatable for key responders to any type of incident. On call rotations are predictable.	MTTA is optimized and repeatable for Vendors, customers or any other allied responders respond as expected.	After Action Reviews fully integrated for Q/A & Q/I of the response team and the process

Figure 11.1

Phase 1 – Initial (Reactive)

At this stage, incident response is ad hoc and largely dependent on individual effort. Success or failure depends on the knowledge, experience, and sometimes heroics of whoever happens to be available. There are no standard playbooks, escalation rules, or communication templates. Communication is inconsistent and often delayed, leading to customer frustration and internal confusion. Metrics are either absent or limited to anecdotal tracking. The economic reality: costs are high because downtime is prolonged, labor hours are wasted, and trust is quickly and regularly eroded.

Economically, phase 1 is the company's most expensive phase. Organizations at this stage often experience hidden downtime where teams are consumed by constant firefighting instead of building strategic value. Most organizations do not remain in Phase 1 for long, because the pain is obvious and severe. However, many organizations return to this phase temporarily during periods of rapid growth or leadership turnover.

	Phase 1
Maturity State	**Initial**
Descriptive State	**Reactive**
Incident roles	• No formalized incident response (IR) training • No clear roles and responsibilities for incident management • Incident communications to key stakeholders and customers is not formalized with a recognized role
Processes	No documented process for dispatch, resolver engagement, resolution or After Action Reviews
Engagement	Company does not recognize or support incident response as an entity or discipline. Best efforts are relied upon from individual contributors

Figure 11.2

Phase 2 – Managed (Responsive)

In Phase 2, organizations begin to recognize that ad hoc actions and occasional heroics are no longer sustainable and that some degree of structure is necessary. While processes may still be immature, there is at least a growing awareness that response requires coordination. Rules of engagement begin to emerge, but they are not yet uniformly applied.

In a Phase 2 organization, some basic templates, escalation pathways, and playbooks may exist, but they are often inconsistently applied. Playbooks look like an easy fix, and many inexperienced ICs ask for detailed step by step instructions on how to run an incident by the book. This is a trap because there is no standard incident. Given the velocity of changes to the production environment caused by capacity planning, technology refreshes, code and configuration updates, and security patching, IT environments are not static. Resolution of every possible incident cannot be reduced to a playbook. As a result, two different teams might handle the same type of incident in very different ways—one using a predictable approach, another relying on improvisation.

	Phase 2
Maturity State	**Managed**
Descriptive State	**Responsive**
Incident roles	• Some level of internal onboarding for incident response • Key incident response roles implemented to some level • No clear and efficient linkage between resolution and key stakeholder communication
Processes	• Some level of formalized dispatch process and SLA's for key responders • Monitoring tools integrated into situational awareness • Blameless After Action Reviews may be completed, but not integrated info Q/A & Q/I
Engagement	Mean Time to Assemble (MTTA) is unpredictably. Dispatching tools, process and accountability in place, but inefficient or outdated or inconsistently used.

Figure 11.3

Key characteristics of Phase 2 maturity include:

- Operations engagement increases, but roles remain blurred. Staff are more consistently brought into the response process, yet responsibilities are rarely defined with precision.
- Accountability is uneven. Some managers or incident commanders drive strong performance, while others allow drift and inconsistency. This creates variability in outcomes.
- Customer-facing communication improves, but only marginally. Updates may occur more frequently than in Phase 1, but they lack clarity, timeliness, or consistency. The net result is that customers may notice improvement, but not enough to feel confident in long-term reliability.

This is the phase where many organizations plateau. We have polled hundreds of responders from a wide variety of companies around the world. Before adopting our incident response philosophy and training, the average OM score from that cohort is 2.5 out of 5! They have moved beyond the chaos of Phase 1, but their processes remain fragile. There is often a false sense of security—leaders perceive the organization as good enough because they no longer experience total disorder. Yet in reality, they remain far from predictable or economically optimized.

Responsonomics emphasizes that Phase 2 is dangerous precisely because of this complacency. The absence of chaos is mistaken for operational maturity, when in fact, the organization is still reacting rather than responding in a predictable way. They may be faster at assembling teams or issuing updates, but still lack the systemic foresight, economic awareness, and cultural discipline required to reduce true cost and price.

In our case study of RayneCloud, the team self-assessed between Phase 2 and Phase 3, which is a common reflection point for organizations seeking to mature further. At this juncture, leadership often recognizes that improvement requires more than patchwork fixes—it requires the deliberate development of defined roles and responsibilities, enforceable rules of engagement, and economic framing, all of which transforms incident management from firefighting into a predictable, cost-aware business function.

Phase 3 – Defined (Predictable)

Phase 3 marks a critical point in an organization's incident management maturity. At this stage, incident response transitions from being reactive and variable to becoming disciplined, structured, and repeatable. In the BlackRock Operational Maturity Model, this phase is referred to as Defined—and in Responsonomics terms, it's where operational consistency begins to unlock measurable economic efficiency. Typically, at this phase, Rules of Engagement (ROE) are not just written—they are actively socialized, understood, and used. This codification eliminates ambiguity during high-pressure incidents. Escalation pathways are understood and well-practiced, removing confusion about who should be involved and when. Time previously spent navigating organizational silos is now reclaimed. Incident Commanders are designated, trained, and empowered. Their presence creates a consistent decision-making structure and limits duplicated or counterproductive actions.

	Phase 3
Maturity State	**Defined**
Descriptive State	**Predictable**
Incident roles	• Command staff functions recognized, implemented and supported by leadership • All key responders and SME's trained to a consistent standard • Regular and predictable briefing/comms cadence
Processes	• On-call rotations predictable and key responders assemble quickly • Playbooks and Standard Procedures developed • Accountability for performance and responder duties is clear to all
Engagement	MTTA is optimized and repeatable for key responders to any type of incident. On-call rotations are predictable.

Figure 11.4

The result: Incidents begin to follow predictable operational patterns. While surprises still happen, responders no longer scramble—they respond within a familiar and well-rehearsed process and under supportive executive leadership.

From an economic perspective, Phase 3 is where incident labor costs begin to flatten. In earlier phases, incidents are plagued by inefficiencies: delayed response, misrouted escalations, unclear authority, duplicated effort. All of these consume time—the fundamental labor cost equation in Responsonomics.

In Phase 3, labor utilization becomes more efficient: Responders are paged more selectively and contribute more purposefully. Coordination overhead drops and communication follows a predictable cadence and structure, minimizing unnecessary status meetings or redundant updates. Fewer responders are needed per incident, because roles and responsibilities are clearly assigned, reducing swarming and over-escalation

Example: A SEV-1 incident that previously required 10 responders for 2 hours (20 labor-hours) can now be handled by 6 responders for 1.5 hours (9 labor-hours). At a $150/hour blended rate, that's a reduction from $3,000 → $1,350 per incident, a 55% savings.

Strategic Benefits at Phase 3

- Business leaders gain confidence in the incident response team.
- Budgets and staffing plans can be based on forecasted incident loads, not guesswork.
- Resilience investments (e.g., better tooling, training, or automation) can now be tied to measurable ROI.

In effect, Phase 3 is where incident response evolves into a managed economic system—ready to feed real-time data into strategic planning and organizational decision-making.

Phase 4 – Data Driven (Repeatable/Scalable)

Phase 4 marks the beginning of true operational maturity. In the BlackRock Operational Maturity Model, this phase is where organizations move into data-driven optimization. From a Responsonomics perspective, this is where the organization becomes capable of producing financial insight, operational predictability, and strategic ROI. At this level, key operational and economic metrics are tracked consistently and used intentionally for real process improvements. Unlike earlier phases where data may be collected passively (or only for compliance/reporting), Phase 4 organizations actively interrogate their data to:

	Phase 4
Maturity State	**Data Driven**
Descriptive State	**Repeatable/Scalable**
Incident roles	• Ongoing training and exercises for all key responders • Large scale incident management in place where applicable <<Unified Command>> • Dedicated communications function assigned
Processes	• Full support of the end to end IR process from senior leadership • All responders accountable for performance
Engagement	MTTA is optimized and repeatable for vendors, customers or any other allied responders

Figure 11.5

- Identify and remove response bottlenecks.
- Optimize escalation paths.
- Predict incident severity patterns.
- Allocate labor strategically.
- Target high-yield automation opportunities.

Also, metrics and modeling capability that matured in earlier phases begins to produce forward-looking insights. Teams can model:

- Cost curves across incident types.

- ROI from proposed staffing adjustments.
- Financial risk from change failure.
- OC of recurring incidents.
- Savings potential from observability tuning.

This supports strategic capital planning—not just for operations, but for board-level investment decisions. Engineering leaders can now enter financial discussions with data-backed forecasts and modeled ROI cases that rival marketing, sales, or product teams in rigor. This is the Responsonomics turning point: when incident response becomes a value conversation, not a cost justification. Phase 4 organizations also develop internal muscle for using data responsibly and repetitively. This produces a culture of repeatability and scalability. New hires, new services, or higher volume can be absorbed without introducing chaos, because the systems are designed to scale with integrity.

Phase 5 – Optimized (Sustainable)

In Phase 5, organizations reach a level where incident response becomes not just operationally effective, but strategically embedded, economically optimized, and culturally reinforced. This phase represents the full realization of Responsonomics principles. At this level, incident response is no longer viewed as a necessary cost center or a

	Phase 5
Maturity State	**Optimized**
Descriptive State	**Sustainable**
Incident roles	• Clear plans to recruit and replace team members • IR team may reach out to customers/key clients/ other business units to assist with building joint response capabilities
Processes	IR is accepted as an integral part of defending the business against financial loss, reputational risk, and loss of customer trust
Engagement	After Action Reviews fully integrated for Q/A & Q/I of the response team and the process

Figure 11.6

reactive function—it is treated as a strategic capability. It is woven into the organizational fabric much like a city's fire department: Always ready, deeply trained, and proactively reducing future risk. Incidents are anticipated, rehearsed, debriefed, and continuously mined for insight. Economic visibility is the norm, not the exception. Organizations in Phase 5 operate with fully institutionalized feedback loops. Every incident becomes a learning opportunity, not just for technical teams but for the business as a whole. Key feedback mechanisms include:

- Postmortems feeding back into detection and prevention tooling.
- Metrics feeding models, and models feeding strategic investment decisions.
- Cultural alignment feeding behavior, where everyone understands that protecting trust, time, and talent is everyone's job.

This feedback-rich environment transforms every incident into economic data that can shape hiring plans, automation priorities, training needs, and customer communications. In this phase, and to some degree Phase 4, Metrics + Models + Culture = Resilience

This is the point where Responsonomics reaches full maturity. Decisions are driven by economic insight, incident response is predictable and cost-effective, and the entire system behaves like a well-governed economy—one where resources are allocated rationally, costs are minimized, and value is protected.

In short, Phase 5 is the economy of incident response working at full throttle—measuring what matters, modeling what will happen, and maintaining a culture that knows that the stakes are high, even when systems seem calm.

To summarize and illustrate how operational maturity can be assessed, consider our hypothetical company, Northridge Systems:

A mid-sized SaaS provider, Northridge prides itself on keeping a lean engineering team and moving quickly with product releases. But, after three months of repeated SEV 1 customer-facing outages, the executive team convened an internal review to better understand what was going wrong with their incident response program.

The findings were concerning but not surprising to the executive team:

- There was consistent role confusion. During a recent payment service outage, more than 40 engineers logged into the bridge call within the first hour, yet no one was clearly in charge. Senior engineers were pulled away from project work, while junior staff hesitated to speak up and incident leadership was relegated to the loudest technical voice on the bridge.
- Incident communication was consistently inconsistent. Customers routinely received no official status update for the first 75 minutes of downtime. In that void, account managers improvised their own messaging, leading to inconsistent promises and increased frustration among key clients.
- Executive mistrust was prevalent: In the absence of good actionable information on a regular cadence from the incident response team, executives joined the bridge to see what was (or was not) happening. Their very presence intimidated more junior responders, which stopped forward momentum.
- Escalation protocols were inconsistent. Some teams followed escalation playbooks while others relied on personal networks. The result was duplicated effort and critical delays.
- Cultural misalignment: A recent company cost saving initiative moved some engineering functions to a lower cost labor market. These teams, now residing in a different time zone and staffed with contractors, were mismatched to the US based headquarters and Executive teams, required off hours for engineers to be on call. This caused churn in the team as some engineers were not interested in overnight work times,

disrupting the established cohesion in the corps of Incident Commanders.

When the leadership team benchmarked Northbridge Systems against the Operational Maturity Model, the conclusion was unavoidable. The company sat firmly at Phase 2. The respondents to the Operational Maturity survey also noted that the team had slid backwards in OM due to the churn caused by the relocation of key engineering functions.

Yes, there were some formal processes in place, but the lack of defined ROE, the inconsistent adoption of communication protocols, and the cultural hesitancy around accountability prevented the organization from progressing further. This is also an illustration of how changes in business practices have a direct and indirect influence on the efficiency of incident response.

The economic impact was clear. Northbridge had spent over $400,000 in wasted engineering hours in the last quarter alone, not including OC from delayed features. More troubling was the erosion of customer trust, as churn risk climbed after each poorly managed outage.

The leadership team left the review with two realizations. First, they could not buy their way out of the problem with yet another tool or monitoring platform—progress required cultural alignment and programmatic investment. Second, unless they moved decisively toward Phase 3 (Defined) the hidden costs of incidents would continue to undermine their growth strategy.

As you can see, turning findings into actionable, data driven conclusions, allowing a company to point themselves down the path of improvement in a deliberate, thoughtful way, is a much more compelling approach to investing in Operational Maturity initiatives.

11 BUILDING A CULTURE OF RESPONSE

Culture eats operations every day of the week

In the final analysis, technology, staffing, and process improvements can only go so far. Culture ultimately determines whether an organization achieves operational maturity or accidently finds specific marginal improvements. A culture that embraces readiness, preparedness, discipline, leadership, accountability and shared ownership creates the foundation for world-class incident response. Conversely, cultures marked by siloed attitudes, resistance to change, or complacency will struggle to reach the highest levels of maturity—no matter how talented their engineers or advanced their tools. Although cost certainly plays a part in implementing change, company culture is often the hidden barrier to the adoption of any change initiative. We've partnered with over 125 companies in 71 cities, training over 7,000 responders in the time Blackrock 3 Partners has been in existence. Based on this experience we can offer some common warning signs indicating that a Responsonomics based change initiative may be challenging. These obstacles are not purely technical. New tools and technology, changes in staffing or company reorganization may temporarily mask organizational weaknesses, but cultural gaps always surface under pressure. Without cultural acceptance, even the best processes and tools cannot be sustained. The warning signs shown below signal cultural misalignment of some kind, somewhere in the organization. These are the visible indicators, but beneath them lie cultural indicators that reveal whether the organization is prepared for transformation. If any of these exist in your company, any change initiative may be difficult.

- Business units and/or functional teams and/or individual labor resources push back on on-call responsibilities, creating gaps in coverage.
- Employees avoid or decline incident response outside "normal hours," slowing detection and resolution.
- Teams operate with their own rules, naming conventions, and escalation practices, leading to inconsistency and confusion.
- Poorly optimized platforms (e.g., paging, observability) remain unchanged because cultural rigor is missing.
- Responders or business units view incidents as someone else's problem.
- Incidents stall while ownership is debated rather than resolved.
- Employees fear blame in post-incident reviews.
- Teams hide mistakes or avoid transparent reporting.
- Executives treat incident management as a necessary evil, not a strategic capability.
- Senior leaders consistently skip or downplay postmortems.
- Frequent incidents are accepted as "just the way it is."
- Past incident outcomes drive a belief that the current system works, even if it relies on heroics or luck.
- Investment in resilience is dismissed because "We are just too busy."
- Certain groups shoulder the burden of incident response, while others avoid participation.
- Hero responders are celebrated, while systemic problems are unrecognized or ignored.

Each of these cultural warning signs points to economic consequences:

- Delays caused by unclear ownership → higher labor and OC.
- Hidden mistakes → repeat incidents and wasted capacity.
- Leadership apathy → underinvestment in prevention, resulting in escalating costs.
- Normalization of deviance → silent erosion of trust with customers and employees.

In short, cultural misalignment creates invisible liabilities that silently accumulate until a high-severity incident exposes them in grand fashion for all to see. At the end of the day, Responsonomics is not just an economic framework; it is a cultural mandate. An organization is ripe for change when it can recognize these warning signs, acknowledge cultural debt, and commit to addressing it openly. Without cultural readiness, even the best technical and financial models will fail to take root.

To close the book, Responsonomics challenges readers to view culture change as a strategic investment. This requires a mandate and commitment from leadership that incident response is critical to protect the value of the business every single day. The company must support and promote continuous skill development and simulations to keep incident teams sharp. Importantly, senior executives must model respect for all responders and escalation protocols. On the flip side, the response team must consistently perform at a high level to earn and maintain that trust because at its core, incident response is about trust and protecting value. Internal employees trust that the incident response team represents the best chance at a good outcome when high severity incidents threaten the business. Customers trust that when systems fail, the organization will respond with speed, rigor, expertise and care. Technology and processes can support that trust, but only culture reinforces and guarantees it.

Responsonomics teaches us to look beyond blinking dashboards and SEV-level timelines. It reframes the work of incident response from a reactive technical chore into a mature economic discipline—one that accounts for cost, value, time, and trust. Responsonomics gives provides the tools to surface and calculate all that. In mature organizations, incident response is managed not as a side-function of engineering, but as a frontline business resilience capability. Responsonomics helps technical leaders speak the language of the boardroom: dollars, risk, and return.

The most mature organizations are not just technically advanced; they are culturally aligned. They refuse to accept "good enough" when customer trust is at stake. Economic maturity without cultural commitment is fragile. But when the two are aligned, operational excellence becomes visible, predictable and enduring.

Glossary

After Action Review
An after action review (AAR) is a technique for improving process and execution by analyzing the intended outcome and actual outcome of an action and identifying practices to sustain, and practices to improve or initiate, and then practicing those changes at the next iteration of the action[1][2] AARs in the formal sense were originally developed by the U.S. Army.
Source: https://en.wikipedia.org/wiki/After-action_review

Business Impact
Business impact is anything that changes the business's current processes and operational activities. It can be positive or negative, depending on the circumstances.
Source: https://cio-wiki.org/wiki/Business_Impact#:~:text=A%20business%20impact%20is%20anything,the%20different%20impacts%20of%20disruptions.

Capacity Leakage
Capacity Leakage refers to the loss or wastage of a project team's potential, which can manifest in several ways. It typically involves inefficient use of time, skills, and resources, often leading to reduced productivity and quality. Some common forms of Capacity Leakage include:

1. Lack of Knowledge or Skills: When team members lack the necessary expertise for their assigned tasks, they may struggle to complete work efficiently, leading to delays and the need for additional oversight or corrections.

2. Misallocation of Resources: Assigning tasks to individuals without considering their strengths or workload can result in inefficiencies and bottlenecks, as well as underutilization of available skills.

3. Inappropriate Participation in Activities: Involving team members in activities where they cannot effectively contribute can waste valuable time and resources, detracting from tasks where their input is more valuable.

4. Excessive Time Spent on Correcting Deviations: A significant amount of project time can be consumed by addressing deviations from plans or correcting errors, especially if these issues stem from miscommunication, unclear instructions, or lack of training.

Source: https://www.linkedin.com/pulse/why-isnt-working-you-may-experiencing-capacity-bogado-nocera-7uiof/

Cost
Cost includes the direct and indirect resource expenditure [predominately the human element] required to investigate, lead, manage, and resolve an incident. It is a measurable expression of the economic footprint that a disruption leaves on the organization.
Source: Authors

Coordination Debt
Coordination debt, in business, refers to the friction, delays, and inefficiencies from teams failing to align efforts, similar to technical debt but for collaboration, leading to stress, slow decisions, and wasted work, often due to silos or unclear ownership.
Source: Google Search AI

Data
Data are a collection of discrete or continuous values that convey information, describing the quantity, quality, fact, statistics, other basic units of meaning, or simply sequences of symbols that may be further interpreted formally.
Source: https://en.wikipedia.org/wiki/Data

Demand
In traditional economics, demand refers to the quantity of a good or service that consumers are willing to purchase at a given price. In Respononomics, demand takes on a unique operational meaning: it represents the human and technology resources required to efficiently resolve incidents [Supply].
Source: Authors

Direct Cost
Any cost that's involved in producing a good, even if it's only a portion of the cost that's allocated to the production facility, are included as direct costs. Some examples of direct costs are listed below: direct labor; direct materials; manufacturing supplies; wages for the production staff; fuel or power consumption
A direct cost is directly associated with the production of specific goods or services and can be traced to a cost object like a product, service, or department.
Source: https://www.investopedia.com/terms/d/directcost.asp

Dispatch
To send (someone or something) quickly to a particular place for a particular purpose.
Source: https://www.britannica.com/dictionary/dispatch

Emergent Fragility
Emergent fragility describes a system's vulnerability, not inherent in its parts, but arising from complex interactions and dependencies, making it susceptible to shocks like economic crises or conflicts, characterized by poor coping mechanisms and cascading failures, as seen in fragile states or complex projects.
Source: Source: Google Search AI

Escalation
Escalation is the structured process of identifying, routing, and resolving issues that exceed the capability or authority of the initial

contact point, ensuring complex problems are handled by the right people at the right time for swift resolution and customer satisfaction.
Source: Google Search AI

Event
In ITIL, an event is defined as "any detectable or discernible occurrence that has significance for the management of the IT infrastructure or the delivery of an IT service," essentially a change of state in a Configuration Item (CI) or service that monitoring tools pick up, ranging from routine logs to potential issues like a server crash or a security alert.
Source: Source: Google Search AI

Friction Cost
The economic impact of inefficiencies, delays, communication breakdowns, and cultural barriers that slow incident response, increase labor expenditure, elevate OC, and amplify the overall price of an incident.
Source: Authors
For more information on the finer points of Incident Response, check out our other book on that topic: Incident Management for Operations by Rob Schnepp, Ron Vidal, and Chris Hawley. O'Reilly Media, 2017.

- The following methods are useful tools when analyzing incidents: Using the 5 Why's, ask who, what, when, where and why? Next, following the practice of Dr. Sakichi Toyoda, ask "Why?" five consecutive times, each time asking the current why on the answer to the previous why. In each case, those finding should be referred to the owners of the relevant technology production systems, for them to take corrective action so a similar incident does not occur again.

Incident
An unplanned occurrence (natural, technological, or human-made) that necessitates a response to protect life, property, critical infrastructure,

or environment.
Source: https://www.directives.doe.gov/terms_definitions/incident

Incident Lifecycle
Every incident goes through a series of stages before it's resolved. It starts when an issue is reported, moves through diagnosis and resolution, and ends once the service is fully restored. That sequence is known as the Incident Management lifecycle.
Source: https://blog.invgate.com/incident-management-lifecycle

Incident Management
The activities of an organization to identify, analyze, and correct hazards to prevent a future re-occurrence.
Source: https://en.wikipedia.org/wiki/Incident_management

Incident Response
The process where a disruption is detected, people are notified and active problem solving occurs to restore service. Incident Response is a subset of Incident Management.
Source: Authors

Indirect Cost
Indirect costs are costs that are not directly accountable to a cost object (such as a particular project, facility, function or product). Like direct costs, indirect costs may be either fixed or variable. Indirect costs include administration, personnel and security costs. These are those costs which are not directly related to production. Some indirect costs may be overhead, but other overhead costs can be directly attributed to a project and are direct costs. There are two types of indirect costs. One is the fixed indirect costs, which are unchanged for a particular project or company, like transportation of labor to the working site, building temporary roads, etc. The other are recurring indirect costs, which repeat for a particular company, like maintenance of records or the payment of salaries.
Source: https://en.wikipedia.org/wiki/Indirect_costs

Labor
Workers considered as a group and an economic group comprising those who do manual labor or who work for wages.
Source: https://www.merriam-webster.com/dictionary/labor#:~:text=bor%20%CB%88l%C4%81%2Db%C9%99r-,1,performed%20by%20workers%20for%20wages

Lagging Indicators
A lagging indicator is a measurable factor that reflects changes or trends after they've already happened, providing historical data to confirm past performance rather than predict future outcomes, like unemployment rates confirming an economic shift or sales revenue showing the result of past efforts.
Source: Google Search AI

Leading Indicators
A leading indicator is a measurable metric that changes before broader economic or business trends, providing a forward-looking signal to help forecast future conditions, risks, or success, allowing for proactive adjustments.
Google Search AI

Mean Time to Assemble (MTTA)
MTTA is the average time it takes from when an alert is triggered to when the dispatched SMEs assemble on the incident command channel. This metric is useful for tracking your team's responsiveness and your alert system's effectiveness.
Source: Authors

Mean Time to Detect (MTTD)
In IT and systems resolution, Mean Time to Detect (MTTD) is to the average time it takes your teams and systems to detect a fault.
Source: https://www.splunk.com/en_us/blog/learn/mean-time-to-detect-mttd.html

Mean Time to Resolve/Repair/Restore/Remediate (MTTR)
Mean time to repair (MTTR) is a basic measure of the maintainability of repairable items. It represents the average time required to repair a failed component or device.
Source: https://en.wikipedia.org/wiki/Mean_time_to_repair

Measurement
The process of associating numbers with physical quantities and phenomena.
Source: https://www.britannica.com/technology/measurement Metric
Measures of quantitative assessment commonly used for assessing, comparing and tracking performance or production.
Source: https://www.investopedia.com/terms/m/metrics.asp

MIM
Major Incident Management (MIM) is the structured process for coordinating a rapid, cross-functional response to severe service disruptions (major incidents) that significantly impact business operations, revenue, or customer experience, aiming to restore critical services ASAP, communicate effectively, and minimize overall business damage.
Source: Google Search AI

Model
An economic model is a simplified version of reality that allows us to observe, understand, and make predictions about economic behavior. The purpose of a model is to take a complex, real-world situation and pare it down to the essentials. If designed well, a model can give the analyst a better understanding of the situation and any related problems.
Source: https://courses.lumenlearning.com/wm-microeconomics/chapter/economic-models/

Notification
Something that gives official information to someone.
Source: https://www.britannica.com/dictionary/notification
Observability
In IT and cloud computing, observability is the ability to measure a system's current state based on the data it generates, recorded as logs, metrics, and traces:

- Logs record the details of an event
- Metrics capture the numeric measurements used to quantify the performance and health of services
- Traces track how services connect from end to end in response to requests

Source: https://www.dynatrace.com/news/blog/what-is-observability-2/

Operational Maturity
Operational maturity defines how consistently, efficiently, and effectively an organization manages its processes, people, and technology to deliver reliable services, handle risks, and adapt, moving from reactive fixes to proactive, strategic operations, often measured by metrics like incident resolution time and service consistency, with high maturity enabling scalability and growth.
Source: Google Search AI

Opportunity Cost
The loss of potential gain from other alternatives when one alternative is chosen.
Source: New Oxford American dictionary

Post Mortems
An analysis or discussion of an event after it is over.
Source: https://www.merriam-webster.com/dictionary/postmortem

Price
In microeconomics, price reflects what a buyer is willing to pay. In Responsonomics, Price is what the company charges itself financially, and perhaps pays emotionally, experientially, and reputationally when a service fails—and how that financial impact and company memory shapes future behavior.
Source: Authors

Priority Levels
Information Technology Infrastructure Library (ITIL) says that Priority should be a product of the Impact/Urgency matrix. ISO/IEC 20000 agrees with that in 8.1 Incident and service request management.
It is customary that Priority has four to five levels, and is marked with the numbers 1-4 or 1-5, where "1" is the highest and "5" is the lowest priority. It can also be marked by letters ABCD or ABCDE, with A being the highest priority.
If one or more services are down, the number can be determined from CMDB data or the service catalog. If more than one service is impacted, parameters for the higher urgency service will be taken into account.
Source: https://advisera.com/20000academy/knowledgebase/incident-classification/

Responsonomics
A new and different way to look at Incident Management and incidents. Responsonomics overlays well-established economic concepts onto Incident Management and Incident Response, reframing service outages, performance degradations, and security events not just as technology interruptions—but also as novel microeconomic episodes that negatively impact labor, productivity, company reputation, brand value, investor confidence and organizational efficiency.
Source: Authors

ROI
Return on investment (ROI) is a financial ratio that measures the profit generated by an investment relative to its cost. Key factors influencing ROI include the initial investment amount, ongoing maintenance costs, and the cash flow generated by the investment. To calculate ROI, the return of an investment is divided by the cost of the investment. The result is expressed as a percentage or a ratio.
Source:
https://www.investopedia.com/terms/r/returnoninvestment.asp

Root Cause Analysis
In science and reliability engineering, root-cause analysis (RCA) is a method of problem solving used for identifying the root causes of faults or problems.[1] It is widely used in IT operations, manufacturing, telecommunications, industrial process control, accident analysis (e.g., in aviation,[2] rail transport, or nuclear plants), medical diagnosis, the healthcare industry (e.g., for epidemiology). Root-cause analysis is a form of inductive inference (first create a theory, or root, based on empirical evidence, or causes) and deductive inference (test the theory, i.e., the underlying causal mechanisms, with empirical data).
RCA can be decomposed into four steps:

1. Identify and describe the problem clearly
2. Establish a timeline from the normal situation until the problem occurrence
3. Distinguish between the root-cause and other causal factors (e.g., via event correlation)
4. Establish a causal graph between the root-cause and the problem.

RCA generally serves as input to a remediation process whereby corrective actions are taken to prevent the problem from recurring. The name of this process varies between application domains. According to ISO/IEC 31010, RCA may include these

techniques: five whys, failure mode and effects analysis (FMEA), fault tree analysis, Ishikawa diagrams, and Pareto analysis.
Source: https://en.wikipedia.org/wiki/Root-cause_analysis

Service Level Agreement (SLA)
A service-level agreement (SLA) is an agreement, typically a binding contract, between a service provider and a customer that define particular aspects of the service – such as quality of service, availability, responsibilities – as terms of contract.[1] SLAs are commonly used in the technology industry, and often contain specific service-level objectives for defined service level indicators, such as mean time between failures, mean time to repair, mean time to recovery, network throughput, jitter, uptime, first-call resolution, and turnaround time.
Source: https://en.wikipedia.org/wiki/Service-level_agreement

Service Level Objective (SLO)
SLOs are agreed upon as a means of measuring the performance of the service provider and are outlined as a way of avoiding disputes between the two parties based on misunderstanding.
Source: https://en.wikipedia.org/wiki/Service-level_objective

Severity Levels
A standardized framework to triage and prioritize incidents based on BI and urgency, ranging from critical outages to minor or informational events. There's no universal definition for severity levels. How you define them depends on what's important to your organization and your users. For some companies, only three levels make sense. For others, dividing incidents into five may be a better idea.
Source: https://www.splunk.com/en_us/blog/learn/incident-severity-levels.html

SME
A subject-matter expert (SME) is a person who has accumulated great expertise in a particular field or topic, which expertise is reflected by the person's degree, licensure, and/or years' occupational experience in

the subject.
Source: https://en.wikipedia.org/wiki/Subject-matter_expert

Specific Marginal Improvements
Small changes and improvements to the overall process of incident management and response. Deliberate and connected process improvements all along the continuum of response will have a significant positive overall impact in incident resolution. Take every opportunity to continuously make specific marginal improvements, because small changes made every day add up to significant improvements over a long period of time. Don't try to eat the elephant of 'getting better' in one bite – break it up into many smaller bites!
Source: Authors

Supply
In conventional economics, supply refers to the specific amount of goods and services available to meet demand. There are more specific components to the traditional definition but for our purpose here, it is sufficient to make the point. However, in the context of Respononomics, supply represents the number and type of incidents and events that enter or have the potential to enter the incident response process.
Source: Authors

Tabletop Exercise
A discussion-based exercise where personnel with roles and responsibilities in a particular IT plan meet in a classroom setting or in breakout groups to validate the content of the plan by discussing their roles during an emergency and their responses to a particular emergency situation. A facilitator initiates the discussion by presenting a scenario and asking questions based on the scenario.
Source: https://csrc.nist.gov/glossary/term/tabletop_exercise

Total Cost of Automation
The real cost of automation includes everything from software

integration to maintenance, training, and ongoing support.
https://brightpick.ai/resources/the-definitive-guide-to-total-cost-of-ownership-for-warehouse-automation/

Trust Erosion Price (TEP)
Trust erosion refers to the gradual decline of confidence and belief that stakeholders have in an organization, often as a result of negative events, poor communication, or unethical behavior. This phenomenon can significantly impact an organization's reputation, stakeholder relationships, and overall effectiveness, especially during crises or issues that challenge public perception.
Source: https://fiveable.me/key-terms/introduction-public-relations/trust-erosion

Unified Command
Unified Command is a team effort process, allowing all agencies with geographical or functional responsibility for an incident, to assign an Incident Commander to a Unified Command organization. The Unified Command then establishes a common set of incident objectives and strategies that all can subscribe to. This is accomplished without losing or giving up agency authority, responsibility or accountability.
Source:
https://www.usda.gov/sites/default/files/documents/ICS300Lesson04.pdf

Bibliography

- Atlassian. Incident Management Handbook. https://www.atlassian.com/incident-management
- Austin, R. D. Measuring and Managing Performance in Organizations. Dorset House, 1996.
- Basiri, Anousheh, et al. "Chaos Engineering." Communications of the ACM, vol. 62, no. 1, 2019, pp. 44–49.
- Beyer, Betsy, Chris Jones, Jennifer Petoff, and Niall Richard Murphy. Site Reliability Engineering: How Google Runs Production Systems. O'Reilly Media, 2016.
- Beyer, Betsy, et al. The Site Reliability Workbook: Practical Ways to Implement SRE. O'Reilly Media, 2018.
- Coase, Ronald. "The Nature of the Firm." Economica, 1937.
- Drucker, Peter. Management: Tasks, Responsibilities, Practices. Harper Business, 1974.
- Forsgren, Nicole, Jez Humble, and Gene Kim. Accelerate: The Science of Lean Software and DevOps. IT Revolution Press, 2018.
- Gartner. (2023). "Modern SLA Management for Enterprise SaaS Providers." Gartner Research Note ID G00771445.
- Google SRE Book. Managing Incidents in Large Systems.
- Kahneman, Daniel. Thinking, Fast and Slow. Farrar, Straus and Giroux, 2011.
- Kaplan, R. S., & Norton, D. P. The Balanced Scorecard: Translating Strategy into Action. Harvard Business School Press, 1996.
- Kaplan, R. S., & Norton, D. P. Strategy Maps: Converting Intangible Assets into Tangible Outcomes. Harvard Business School Press, 2004.
- Kim, Gene, Kevin Behr, and George Spafford. The Phoenix Project: A Novel About IT, DevOps, and Helping Your Business Win. IT Revolution Press, 2013.
- Kim, Gene, Jez Humble, Patrick Debois, and John Willis. The DevOps Handbook. IT Revolution Press, 2016.

- Marr, Bernard. Key Performance Indicators: The 75 Measures Every Manager Needs to Know. Pearson Education, 2015.
- Murphy-Hill, E., Zimmermann, T., & Nagappan, N. "The Hidden Costs of Disruption: Behavioral Responses to Outages in High-Traffic Systems." Microsoft Research, 2019.
- Woods, David D. "Essential Characteristics of Resilience." In Resilience Engineering Concepts and Precepts, edited by Erik Hollnagel, David D. Woods, and Nancy Leveson. Ashgate, 2006.

Industry Reports & White Papers

- APM Digest. Best Practices in Signal Validation. https://www.apmdigest.com/network-observability-reduce-incident-detection
- Datadog. The State of Alert Fatigue in SRE Teams. https://www.datadoghq.com/blog/alert-fatigue/
- DevOps.com. Three Strategies for Reducing MTTD and MTTR. https://devops.com/three-strategies-for-reducing-mttd-and-mttr-as-outage-costs-spiral
- Honeycomb.io. Observability-Driven Development. https://www.honeycomb.io/blog/observability-driven-development
- New Relic. State of Observability Report 2024. https://newrelic.com/resources/report/observability-forecast/2024/state-of-observability/outages-downtime-cost
- PagerDuty. Incident Response Documentation. https://response.pagerduty.com/
- PagerDuty. How to Improve Alert Quality and Reduce Burnout. https://www.pagerduty.com/resources
- PagerDuty. Automation ROI Calculator. (2022). https://www.pagerduty.com/resources

Technical Blogs & Thought Leadership

- Allspaw, John. "Automation Should Be Boring." Adaptive Capacity Labs Blog, 2020.
- Allspaw, John. "How Complex Systems Fail." Adaptive Capacity Labs, 2012. https://how.complexsystems.fail
- APM Digest. Network Observability & Incident Detection. https://www.apmdigest.com/network-observability-reduce-incident-detection
- Honeycomb.io. (2021). Incident Management Maturity Curve.
- Majors, Charity. "Observability ≠ Monitoring." Honeycomb.io Blog, 2018. https://www.honeycomb.io/blog/observability-is-not-monitoring
- Motadata. Reduce IT Cost Through Observability. https://www.motadata.com/blog/reduce-it-cost-through-observability
- Scheirer, Jeff. "MTTR Is Dead—What You Should Measure Instead." PagerDuty Blog, 2020. https://www.pagerduty.com/blog/mttr-is-dead/
- Salesforce. The Hidden Costs of Poor Incident Management. https://salesforce.com/whitepapers
- Uptime Institute. The Cost of Data Center Downtime. https://uptimeinstitute.com

Miscellaneous References

- AWS Architecture Blog. "Using Observability to Drive Resilience." https://aws.amazon.com/blogs/architecture
- NIST Technical Report. Key Practices in Coordinated Incident Response Teams.
- U.S. Bureau of Labor Statistics. "Average Hourly Earnings of Production and Nonsupervisory Employees." https://www.bls.gov

ABOUT THE AUTHORS

Rob Schnepp's incident response career spans 30 years in international public safety as a Special Operations Fire Chief, Incident Commander, consultant and published author. Rob commanded numerous large-scale emergencies for the Alameda County (CA) Fire Department, protecting 500 square miles and two national laboratories in the East Bay of the San Francisco Bay Area. Rob has responded on Unified Command/Incident Management Teams to high profile incidents including: 9/11 World Trade Center; major oil spills; large propane gas explosions; campaign wildland fires and numerous other large scale incidents. Rob planned and directed Red Command at Urban Shield, the largest Homeland Security exercise in the United States, as well as a full scale exercises for the Department of Defense in 12 countries across the globe. He has developed risk assessment, incident management and incident command training for Fortune 500 companies, foreign governments, and US national laboratories.

Ron Vidal's career spans 45 years in senior leadership roles in critical infrastructure, capital markets and incident management. Prior to co-founding Blackrock 3 Partners, Ron's experience includes data centers, metro/long haul/subsea fiber optic networks, high voltage power systems, oil & gas, mergers & acquisitions, capital raising, investor relations and government affairs. Ron was a senior executive at Level 3 Communication (operates 16,000-mile international metro, long haul and transoceanic fiber optic Tier 1 backbone network), UUNet Technologies (largest Tier 1 Internet Service Provider (ISP) when acquired by MFS), MFS Communications (operated 38 metro fiber networks in U.S. and Europe, acquired by Worldcom) and Kiewit (3rd largest U.S. construction contractor). He led technical and operational due diligence teams on over $19B of telecommunications and Internet M&A transactions. He developed and delivered key messaging during

$14B of equity and debt capital raising across 10 international roadshows and 8 trading desks. He led Level 3 Communications in New York City after the 9/11 and served on the New York City Task Force on Network Reliability. Ron has advocated technology public policy to members of the United States Congress, the Federal Communications Commission (FCC). He testified before the House Subcommittee on Telecommunications and the Internet on E-911 and Voice over Internet Protocol (VoIP) policy

www.ingramcontent.com/pod-product-compliance
Lightning Source LLC
LaVergne TN
LVHW022224200926
848481LV00006B/2102
* 9 7 9 8 9 9 3 5 6 5 4 0 8 *